▼ ▼ ▼

SPEAKERS

for Your Home and Automobile

How to Build and Enjoy a Quality Audio System

- Gordon McComb
- Alvis J. Evans
- Eric J. Evans

PROMPT.
PUBLICATIONS

An Imprint of
Howard W. Sams & Company
Indianapolis, Indiana

▼ ▼ ▼

REVISED SECOND EDITION, 1992

PROMPT® Publications is an imprint of Howard W. Sams & Company,
2647 Waterfront Parkway, East Drive, Indianapolis, IN 46214-2041.

This book was originally developed and then published as Building Speaker Systems
by:
    Master Publishing, Inc.
    14 Canyon Creek Village MS31
    Richardson, Texas 75080
    (214) 907-8938

International Standard Book Number: 0-7906-1025-6

Edited by: Charles Battle, Gerald Luecke
Text Design and Artwork by: *Plunk Design, Dallas, TX*
Cover Design by: *Sara Wright*

Acknowledgements
All photographs not credited are either courtesy of Author, Master Publishing, Inc., or
Howard W. Sams & Company.

Eric Evans expresses appreciation to Sound Works, Fort Worth, TX for their assistance.

Printed in the United States of America

9 8 7 6 5 4 3 2 1

▼ ▼ ▼

___

TABLE OF
CONTENTS

PREFACE

Of all the components in a quality audio system, the weakest link could very easily be the speakers. There are two reasons for this. The first is cost. Quality looking and sounding speakers can be very expensive. The second is that speaker specifications and design seem complicated. The average audio system owner doesn't even want to bother to learn about the component that turns out to be the key to fine listening enjoyment.

Speakers for Your Home and Automobile has been written to address both of these reasons. It should make it easy for you to construct inexpensive, quality speaker systems, and, while doing so, to learn the characteristics of a quality speaker system. *Speakers for Your Home and Automobile* not only tells you how to build the systems, but also tells you what components to choose and why. Easy to understand illustrations visually enhance the learning.

The book begins with an introduction to the dynamics of sound and how speakers reproduce sound. It continues with speaker terms, the various types of speakers and points out the characteristics that determine speaker performance.

Speaker performance is useless without an enclosure to make it a quality sound system. The different types of enclosures and the selecting of design factors to obtain certain system characteristics conclude the discussion on speaker and enclosure design and performance.

Then the discussion on building begins. The construction techniques, putting on quality finishes, wiring the speaker into the enclosure, and connecting the speaker system to an amplifier lead up to Chapter 8 which contains the specification plans and parts lists to build four different home speaker systems.

Chapters 9 and 10 then shift to automotive systems. Chapter 9 deals with the overall system and improvements that can be made by replacing factory installed speakers with higher performance units, and/or by adding additional components. Chapter 10 describes specific system installations in coupes or sedans, hatchbacks, pickup trucks, and vans or sport utility vehicles.

In addition, a metric conversion chart from inches is included in Appendix C for those who wish to deal in metric units.

Readers who complete *Speakers for Your Home and Automobile* should understand the role of speakers and their enclosure in a quality sound system, and, if they want, be able to choose and construct quality speaker systems. That was our goal. We hope we have succeeded.

G.McC, A.E., E.E.

INTRODUCTION – OF SOUND AND SPEAKERS

Imagine, for a moment, that you are an E flat quarter note from a clarinet in Mozart's "Serenade for Winds." You've been digitally recorded on compact disc and your turn to make beautiful music is coming up. At almost the speed of light, you are transferred from the surface of the disc and converted into electrical pulses, and are on your way to a stereo amplifier.

For an electrical signal, you have traveled far; still, little of your personality has changed — the excellent sound system you are being played through adds virtually no spurious sounds of its own.

Finally, you reach the speakers of the sound system, and you are converted from electrical signals to sound. Unlike the rest of the stereo system, the loudspeakers aren't designed very well, and as you rush through the air, you realize your personality has changed drastically. You no longer sound like a mellow E flat quarter note from a clarinet, but a tinny rasp like that of a kazoo. All that work — for nothing.

The loudspeakers are the last link in the high-fidelity audio chain, and it is the element that is most often ignored. Stereo enthusiasts routinely spend hundreds — even thousands — of dollars on the latest high-tech compact disc, turntable, cassette, and receiver gear. Purchases are carefully selected, with an eye on technical specifications and performance. But the loudspeakers are given a brief once-over. Specifications are largely ignored, and little thought is given to how the design of the speakers will integrate with the home's acoustic environment.

One of the reasons speaker systems are given little consideration is that good ones are expensive. A pair of quality loudspeakers in a handsome oak or walnut enclosure can easily cost $500, $600, and even more. Another reason is that speaker specifications and design seem complicated, not easy to understand.

This book, *Speakers for Your Home and Automobile* addresses both of these objections — and more. It shows you how to save money by building your own speaker systems. Even if you have little woodworking experience, you can construct your own affordable, quality speaker systems in one or two weekends.

While you're having fun putting together your speaker systems, you'll learn:

- About the different types of speakers — woofer, midrange, and tweeter — and how they are used.
- How to understand speaker specifications and make them work for you.
- How enclosure design affects sound quality.
- How to use the enclosure construction plans presented in this book.
- How to compute size and dimensions for your own custom speaker system designs.
- How to build quality speaker enclosures — from small "bookshelf" speakers to huge floor-standing "towers."
- About the role of the crossover network, L-pad, power fuse, and other electrical speaker system components.
- How to connect speakers to your stereo system.

And there is more. Even if you don't plan on building your own speaker system, this book will help you better understand the role speakers play in modern state-of-the-art hi-fi systems. With the easy-to-understand information presented in this book, you'll learn how to interpret loudspeaker specifications, how to judge the merits of the various speaker designs, and how to spot a good set of speakers.

We have a lot to cover in the pages that follow, so let's begin. The remainder of this chapter introduces you to the dynamics of sound, and how speakers reproduce that sound. Understanding the principles of sound production goes a long way in enriching your appreciation of how speakers work.

THE SCIENCE OF SOUND

On a broad level, the sounds you hear — whether they come from a stereo system or from the person standing next to you — are waves that travel through the air. The sound source causes the air particles to vibrate; that is, to move up and down to propogate the waves from source to receiver. These vibrations propagate out from the source in waves, much like the waves which propagate out from a pebble dropped in water. Your eardrums pick up the air vibration and your inner ears turn it into signals for your brain.

Without a fluid like water or air (yes, to the physicist air is considered a fluid), there can be no sound. Sound travels through many fluids, but we are concerned with how it travels through air. The ways you control the air determine what happens to the sound. This is an important point, as you'll see in the chapters to come. Remember, for speaker systems, sound is the movement of air.

Sound Volume

The intensity, or volume, of sound is determined by the amplitude of the air particle vibrations, as depicted in *Figure 1-1*. Waves that fluctuate in the air a small amount produce little sound, discernible to only those near the sound

SPEAKER   EAR

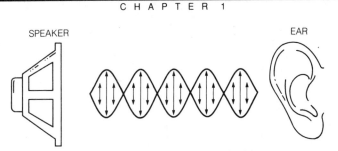

a. Low Volume Waves Through Air

Double arrow indicates up and down motion of air particles that propagate the sound
wave from source to receiver. The particles remain in place; the wave propagates.
Larger amplitude means louder sound.

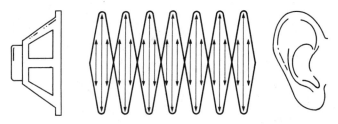

b. High Volume Waves Through Air

**Figure 1-1. The amplitude of low volume waves are small, so the sound is weak.
Conversely, the amplitude of high volume waves are large, so the sound is
strong.**

source. The eardrum isn't vibrated much by low-intensity waves, so the sound
is considered low-level.

Huge waves that thunder about the room produce a large sound. Your
eardrums are vibrated a great deal, and the sound is considered loud. In fact, if
the waves are too large and the sound level is too high, your eardrums may be
damaged.

Technically, sound waves are pressure waves. Vibration of the air
particles in a sound wave cause small changes in local air pressure. The larger
the pressure change, the louder the sound. Sound pressure level (SPL) is
commonly expressed in decibels, a logarithmic unit of measure. The decibel
(dB) is not an absolute measure, but rather indicates a ratio between two sound
levels. Zero dB represents the SPL of a 1000 Hz tone that is just discernible by a
person of normal hearing in an otherwise quiet background. The level of other
sounds is measured or expressed relative to this lowest audible sound level.
Again, for a person of normal hearing, a change in SPL of 10 dB is perceived as
being twice as loud as the original sound.

Figure 1-2 shows a graph depicting various sound intensities and their
relative levels in dB. Note that the measurements are taken at a set distance
from the sound source as shown in the figure. The sound level decreases 50
percent every time you double the distance between the source and your ear.
Studies show the human ear cannot tell a difference of sound level smaller than
1 dB.

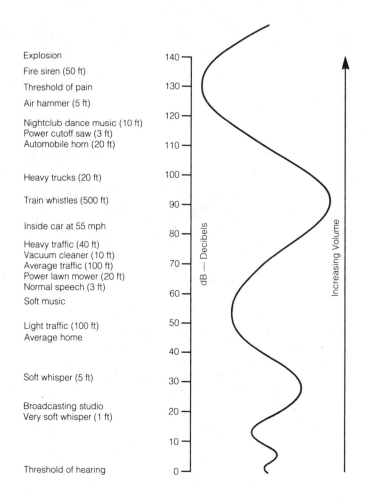

Explosion	140
Fire siren (50 ft)	
Threshold of pain	130
Air hammer (5 ft)	
	120
Nightclub dance music (10 ft)	
Power cutoff saw (3 ft)	
Automobile horn (20 ft)	110
Heavy trucks (20 ft)	100
Train whistles (500 ft)	90
Inside car at 55 mph	80
Heavy traffic (40 ft)	
Vacuum cleaner (10 ft)	
Average traffic (100 ft)	70
Power lawn mower (20 ft)	
Normal speech (3 ft)	
Soft music	60
Light traffic (100 ft)	50
Average home	
	40
Soft whisper (5 ft)	30
Broadcasting studio	
Very soft whisper (1 ft)	20
	10
Threshold of hearing	0

dB — Decibels

Increasing Volume

Figure 1-2. Various sound levels, in decibels (dB). As a comparison, soft music (as heard near the speaker) is approximately 60 dB and requires a amplifier power of less than 0.01 watt. Loud music (again, near the speaker) is about 105 to 115 dB, and requires an amplifier power of 30 to 300 watts. The threshold of pain for most humans is about 130 dB.

A sound level meter, shown in *Figure 1-3*, is often used to accurately measure the level of sound in a given area. The sound level meter is not a necessary item when building and testing your own speakers, but if you are serious about experimenting with sound and loudspeaker designs, you may want to consider purchasing one.

Figure 1-3. A sound level meter. The unit is inexpensive, easy to use, and provides accurate results. *(Courtesy of Radio Shack)*

Sound Frequency

Sound is made up of various frequencies — the rate at which the particles in the air vibrate, or more technically correct, "oscillate." Low frequencies (slow oscillations) produce deep, bass sounds; high frequencies (fast oscillations) produce sounds such as the tinkle of bells or the clash of cymbals. Frequency is measured in hertz (Hz), which is the same as cycles per second. A sound with a 10 Hz frequency oscillates at 10 cycles per second; that is, every second, the air particles move up and down 10 times to propagate the wave at the speed of sound through the air. Humans hear in a relatively narrow frequency band from about 20 Hz to 20,000 Hz. (20,000 Hz is usually expressed as 20 kHz, where k stands for 1,000.) Audio systems, including loudspeakers, are designed to produce sound within this range.

Frequency range is an important aspect of the science of speaker systems. As illustrated in *Figure 1-4*, the human voice and all musical instruments produce sound over the range of frequencies shown.

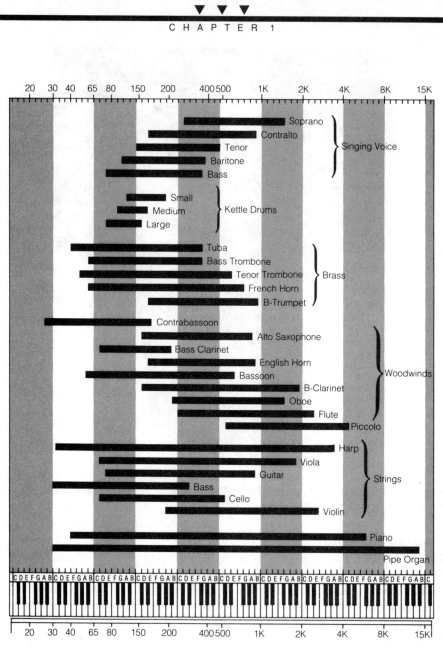

Figure 1-4. The range of human hearing extends from about 10 Hz to 20 kHz. The human voice and musical instruments fall within this range. Note than many instruments, particularly the string variety, span a wide range, and require a speaker system that accurately reproduces these frequencies.

Three Ranges of Sound

The audio spectrum of 20 Hz to 20 kHz can be subdivided into three major categories: low, medium and high as indicated in *Figure 1-5*.

- Low sounds, referred to as bass, are made by bass singing voices and by bass instruments, such as kettle drums, tubas, bassoon, and string bass.
- Middle sounds, called midrange, are made by most singing voices, guitars, and most other musical instruments.
- High sounds, referred to as treble, are made by bells, cymbals, flutes, violins, and crickets trapped in the sound studio.

Most speakers are designed to efficiently reproduce well within one of these three ranges. *Figure 1-5* shows how the common speakers cover the ranges.

- Large speakers, called woofers, reproduce low sounds and some middle sounds.
- Medium-size speakers, called midranges, reproduce middle and some high sounds.
- Small speakers, called tweeters, reproduce all high or treble sounds.

Direct Versus Reflected Sound

Sound can either be direct or reflected. Reflected sound is commonly called an echo, though the term reverberation may be used in audio because reverberation is the addition of all echoes. If you've ever been inside an empty house, in an underground cavern, or any other place with hard, smooth walls, you've heard your footsteps or voice reverberate — they echo. The echoes vary with the enclosure size and the number of surfaces. When you listen to music through a speaker system, you hear a portion of the sound directly from the

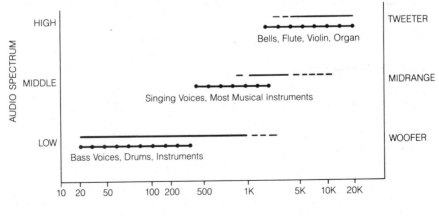

Figure 1-5. The audio range of human hearing, divided into three groups: low, middle, and high. Most speaker systems are designed around these groups. The solid line details the most common range for the respective speaker.

speaker. Other portions reflect from the walls, ceiling, floor, furniture, and other objects. These are echoes. The sum of what you hear at a particular point is the reverberation.

Despite what you may think, echo can be used to advantage if planned carefully. For example, echo increases the intensity of low frequencies and helps speakers reproduce such sounds more efficiently. Echo makes a small speaker sound louder, in a way similar to how your voice seems louder in an enclosed room than it does out in the open. All speaker systems are designed with echo in mind. In the chapters that follow, you'll learn how to use echo (we'll call it reverberation) to make better speakers, and how to place your speakers in the room to take maximum advantage of reflected sound.

HOW SPEAKERS WORK

It's sometimes easier to visualize how speakers work by first looking at how our eardrums work, then reversing the process. For hearing, sound waves enter the ear canal, and vibrate the eardrum. High frequency sounds vibrate the eardrum faster, which in turn makes us hear high pitched sounds. Similarly, low frequency sounds vibrate the eardrum slower.

The eardrum is connected to the inner ear (the cochlea) by a series of miniature bones. Sensors in the inner ear convert the sound waves to electrical signals, and these are routed to the brain.

For speakers, we begin with the electrical signals in the amplifier. A pickup head from a record or tape player sends electrical signals to the amplifier. The amplified electrical signals are applied to terminals on the speaker. The speaker responds to the signals by moving a cloth, plastic or paper cone back and forth. This causes the air around the speaker to pressurize and depressurize, producing sound waves in the air. These sound waves travel through the air and are heard by you. High frequency signals cause the speaker cone to vibrate rapidly and produce high frequency sounds. Low frequency signals produce low frequency sounds because the cone vibrates slowly.

Dynamic Speakers

About 95 percent of all speakers used in stereo systems are the dynamic variety. They use a magnet and a coil of fine wire to produce the movement of the cone, which in turn causes sound. Specialty types include piezoelectric speakers, which are covered later in this chapter and in Chapter 2. The parts of a typical dynamic speaker are shown in *Figure 1-6*.

The frame of the speaker holds everything together and provides the means to fasten the speaker to an enclosure or panel. The frame (also called the basket) has cut-outs in the back so air can freely circulate around the speaker cone. The diameter of the frame can range from one inch to over 15 inches. Generally, the larger the size, the more volume the speaker can produce. Also, larger speakers are predominantly used to reproduce bass sounds; smaller speakers are typically used to reproduce midrange and treble sounds.

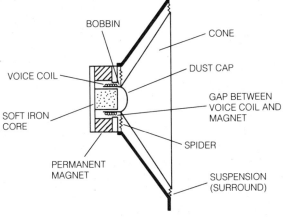

Figure 1-6. The parts of a typical dynamic speaker are identified in this cross-sectional view.

The permanent magnet and voice coil make up the driver of the speaker. The voice coil consists of many turns of fine wire wound on the bobbin. Electrical audio signals from an amplifier are applied to the voice coil. A varying electromagnetic field is produced around the voice coil and this field aids or opposes the magnetic field produced by the permanent magnet. It is as if you are bringing two permanent magnets together, but alternately turning one of them so that sometimes the magnets are pulled together and sometimes are pushed apart. Since the speaker's permanent magnet is fixed in the frame and the voice coil is movable, the voice coil moves as the magnetic fields interact.

As the audio signals are continuously fed to the speaker, the varying electromagnetic field causes the voice coil to move back and forth over the soft iron core of the magnet. As stated previously, the voice coil is wound around a small paper, plastic or metal cylinder called the bobbin. The bobbin, in turn, is attached to the cone. When the coil moves because of the electrical signal applied to it, the bobbin moves the cone and causes it to vibrate. The dust cap placed over the bobbin forms the inside center of the cone and keeps dust and debris from entering into the small gap between the voice coil and the permanent magnet core.

Since the cone moves back and forth as it vibrates, its suspension elements must be flexible. At one end of the cone, the flexible spider attaches the voice coil, bobbin, and inner portion of the cone to the frame. At the other end of the cone is a flexible rubber, foam, or paper element called the suspension (or surround). The two general types of suspension are folded and half-roll. Chapters 2 and 3 help you decide which type of suspension is best for each type of speaker system you build.

Piezoelectric Speakers

The term piezoelectric means pressure electricity. The fundamentals of piezoelectricity were discovered by two French physicists, Jacque and Pierre Curie. They found that certain materials, especially crystals, generate a voltage when they are placed under pressure. The reverse happens when electrical signals are applied to the material; that is, the material vibrates.

Today, piezoelectric elements are commonly used in buzzers and chimes. Another example: The "chirp" in an alarm wristwatch is a piezoelectric ceramic disc responding to an electric current.

Piezoelectric elements also are used in stereo system speaker designs. Since they are most efficient with high frequency sounds, piezoelectric speakers are used almost exclusively as tweeters for hi-fi speaker systems. Of course, you have a choice between using an electrodynamic or piezoelectric tweeter in your stereo system. In later chapters, you'll discover the advantages of both.

BASIC SPEAKER SYSTEM DESIGN

A bare speaker; that is, one not mounted on or in anything, will likely not be operating at maximum efficiency and won't sound very good. To make them more efficient and improve sound quality, all speakers are designed to be placed inside some type of enclosure or cabinet.

Sound waves are emitted from both the front and back of the speaker cone. Pressure waves from the back of a bare speaker can come around the side of the speaker and fill the low pressure area at the front, and vice versa. This results in total or partial cancellation of the generated sound waves. The speaker is far less efficient than it should be. Unintentional sound frequencies, called distortion, can result.

The most rudimentary speaker enclosure design is shown in *Figure 1-7*. The speaker is placed within a sealed enclosure with air trapped inside. When the speaker cone vibrates, the cone is forced back by the pressure of the air sealed inside the cabinet. The enclosure also confines the back wave and prevents its interaction with the front wave, thus avoiding the low frequency sound cancellation discussed above. The result? Speaker performance is greatly improved. In many cases, the speaker enclosure may not be 100 percent airtight, but it's sealed enough to do the job.

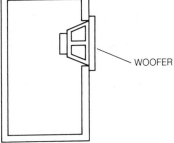

Figure 1-7. The basic sealed-box or acoustic suspension speaker system. Only a woofer is shown; other speaker types may be added.

Because the speaker is cushioned or "suspended" by air inside the enclosure, this design is often referred to as acoustic suspension. If the enclosure is very large ("tower" size), it is often called an infinite baffle.

Another type of speaker enclosure design is called ported reflex, and is shown in *Figure 1-8*. In this design, a tube is built into the cabinet that lets a certain amount of air travel in and out of the enclosure as the speaker cone is in motion. The advantage of the ported reflex enclosure over a sealed box is a deeper bass and higher efficiency. This speaker also is known by other names such as included ducted port, bass reflex, and Helmholtz resonator. Later in this book, you'll learn how to make both acoustic suspension and ported reflex speakers.

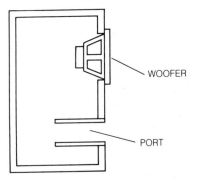

Figure 1-8. The basic ported reflex speaker system. Only a woofer is shown; other speaker types may be added. Note that the port can be a tube that extends into the enclosure, or it can be simply a hole in the side, back, or front of the enclosure. The hole and tube size are critical.

MOVING ON

In this chapter, you learned the basics of sound and how sound is reproduced by a speaker. You learned the anatomy of a speaker and how the parts of the speaker work together. But this is just the tip of the iceberg. There is more to speaker dynamics than what we have covered here. In the next chapter, you'll learn more about speakers, and how speakers made with different materials produce different results.

A CLOSE LOOK
AT SPEAKERS

A lot goes on in the human body to say the word "Hello." The vocal chords vibrate, the throat flexes and tenses, the mouth opens and closes in a variety of shapes, and the tongue flaps up and down like a flag in the wind. Even before the sound leaves the mouth, it's circulated around the head for a little while — to give it resonance. If you happen to have a cold, your sinus cavities and nasal passages may be blocked, and your voice has an irritating rasp to it.

Hi-fi speakers are a lot simpler — and more efficient. As you saw in the last chapter, speakers work by vibrating a cloth, plastic or paper cone. The cone moves the air around the speaker and the resulting air pressure variation causes waves that we hear as sound. In addition to saying "Hello," hi-fi speakers can reproduce the sounds of saxophones, drums, guitars, pianos, violins — you name it.

Though all speakers do the same thing — namely, reproduce sounds from electrical signals — not all speakers are designed alike. There are different sizes, types, and functions of speakers, made with a variety of materials. With the myriad choice of speakers open to you, it's important that you understand the differences between them, and how to apply the various speaker designs to each of your applications.

A QUICK BIT OF TERMINOLOGY

Before continuing with the design of speakers, a number of audio concepts should be explained. By learning the definition of these concepts now, you'll be better able to understand the sections that follow.

Frequency Response

Basically speaking, frequency response is the sonic frequency range that an audio system can evenly and accurately reproduce. It is the variation of the speaker output over a frequency range. It is measured by varying the frequency of an input signal over a frequency range. The input signal amplitude

is held constant. The frequency is specified in hertz (Hz), and extends from about 20 Hz to 20 kHz. A hertz is a signal variation of one cycle per second. The sonic range from 20 Hz to 20 kHz is effectively the range of human hearing, although many people can't hear high frequency sounds. A more reasonable hearing range for most adults is 30 Hz to 15 kHz.

When a series of test tones with a constant amplitude are played through an audio system, every frequency in the 20 Hz to 20 kHz range should have the same output level as the others. When this happens, the frequency response is said to be flat, and is expressed in specifications literature as "20 Hz to 20 kHz, ±0 dB." In reality, no audio system has a perfectly flat frequency response from 20 Hz to 20 kHz. There is some deviation, measured in plus or minus dB, around some midrange value.

The decibel is a basic unit of measurement for signal amplitude. It is a measure of the ratio of the audio system output power compared to the input power. If the audio system amplifies all frequencies the same with constant input power, then output power for all frequencies will be the same. Therefore, the response is flat and the ratio is the same dB value for all frequencies. The higher the dB value, the more power the signal has at the output for a given constant input signal.

Of course, the tones in real music are seldom even; that is, the tones are of varying frequencies at many different input levels. Nevertheless, the audio system should be able to accurately reproduce those tones at its output. It is a matter of what goes in should come out in the same form. If the speaker tends to favor certain frequencies and ignore others, it is said to color the sound. Ideally, the speaker should reproduce all frequencies without coloring.

Frequency response is usually depicted in the form of a graph of output level in dB versus frequency. *Figure 2-1* provides a clear view of frequency response. It shows how the system reproduces midrange frequencies at a fairly constant level, but the level falls below the frequency f_1 and above the frequency f_2. Notice that the frequency response is really a two-prong specification. On the one hand, it's the sonic range that the system can reproduce; and on the other hand, it's the accuracy to which the frequencies can be reproduced relative to the original recording. Notice in *Figure 2-1* that the system output varies plus or minus some dB value around the midrange value. The lower and upper frequencies at which the response is 3 dB below the midrange value are usually identified by f_1 and f_2, respectively.

Dynamic Range

Dynamic range is the difference between the softest and loudest portions in a music selection, and depends largely on the source of the audio signal. For example, compact disc players, which have a theoretical dynamic range of 96 dB, have a practical dynamic range of 85 dB to 90 dB. This is much better than other recording mediums.

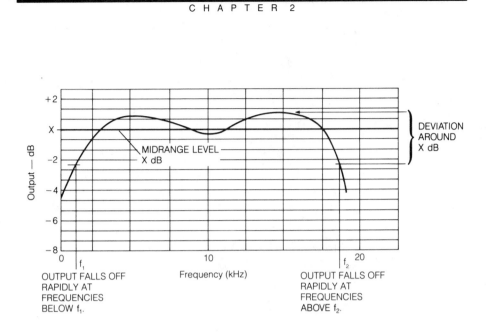

Figure 2-1. A visual representation of frequency response. The curve denotes the upper and lower limits of response; the flatness of the line denotes how well the audio system reproduces the various frequencies over the entire sonic spectrum. The X dB level represents the midrange amplifier output.

These figures are really applicable for laboratory tests; it's unlikely you'll ever find a piece of music with this kind of range — the 1812 Overture included. A 90 dB range means the power of the loudest signal has 10^9 (one billion) times as much power as the softest signal! Even though the signal input and the amplifier have excellent dynamic range, other elements in the audio system, particularly the speakers, restrict the dynamic range. Therefore, if you have an audio system with an overall dynamic range of 60 to 70 dB, it is still a pretty good system.

Sound Dispersion

Sound dispersion is the spreading of sound as it leaves the speaker (see *Figure 2-2*). A narrow dispersion means that the sound does not spread very far. The listener must be directly in line with the axis of the speaker to obtain the best results. A wide dispersion spreads the sound over a larger area, so the effective listening area of the speaker is increased.

Dispersion changes at different frequencies, even with the same speaker. Generally, a speaker is omnidirectional (disperses sound equally in all directions) up to the frequency where the diameter of the cone is equal to the wavelength of the sound. Then it becomes directional for higher frequencies (shorter wavelengths).

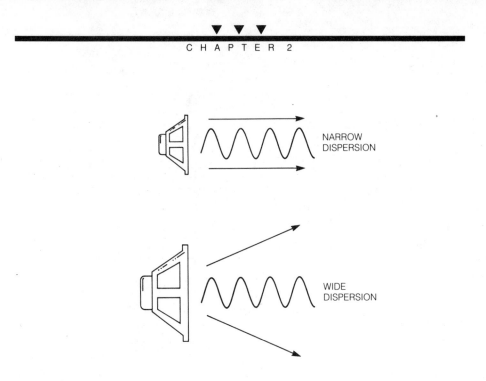

Figure 2-2. Narrow dispersion occurs when sound waves from the speaker do not spread much as they travel through the room. Wide dispersion occurs when the sound waves fan out. Narrow dispersion is a critical factor in high frequency sound. Tweeter and midrange speakers are often designed to overcome narrow dispersion problems.

Since the velocity of sound is 13,200 inches per second (at sea level), the omnidirectional frequency limit can be determined by dividing the velocity by the speaker diameter. The omnidirectional limit for a 15-inch speaker is 880 Hz. For a 12-inch, 8-inch, and 4-inch speaker, the respective frequency limits are 1.1 kHz, 1.65 kHz, and 3.3 kHz. This limit is not exact (speaker materials and cone geometry have an effect, as does the design of the speaker enclosure), so a 4-inch speaker will probably extend to about 4 kHz, and a 2-inch speaker to about 8 kHz.

Sensitivity

The sensitivity of a speaker is a measure of its sound output level for a specified electrical input power. The manufacturer applies a continuous single-tone signal, having a specified electrical power level, to the speaker. The sound output level is then measured at a specific distance (usually one meter). The sensitivity specification is often used to denote the efficiency of the speaker, or how well the speaker converts electric signals into sound. A speaker that puts out a great deal of sound for a given signal input is said to be efficient (but not necessarily better sounding than another speaker).

Damping

Speaker cone motion should faithfully follow the applied electrical signal. Any tendency of the cone to vibrate at frequencies which are not in the input or to continue to vibrate after the signal stops will color the sound and adversely affect its quality. This unwanted cone motion must be damped. Speakers are made with built-in damping mechanisms to tame these unwanted vibrations. Components that damp the action of the speaker are:

- Cone material
- Suspension
- Spider
- Magnet

A stiff cone doesn't vibrate as readily as a pliable cone. Most speakers are made with a relatively stiff cone, one that will maintain its durability over years of use. The suspension and spider hold the cone to the frame. The spider is usually made very supple to readily accommodate the motion of the cone. The suspension can be made hard or soft. The amount of stiffness of the suspension is usually referred to as compliance. A low compliance suspension means it is stiff; high compliance denotes a pliable suspension.

The size and weight of the magnet determines the sound output for a given input signal, but it also determines electrical damping. A small magnet does not provide much force against the voice coil, so the speaker acts like a car with weak shock absorbers. Damping can be tightened by increasing the size and weight of the magnet.

Damping is also determined, as you will see later in this chapter and in Chapter 3, by the size and design of the speaker enclosure. Ideally, the enclosure complements the damping characteristics of the speaker, working with the design of the speaker to produce clear and natural sound.

Sound Distortion

Any sound that you hear coming out of your hi-fi speakers that wasn't originally recorded is deemed distortion. Some distortion is created in the electronic circuitry, including the tuner, amplifier, phonograph, and cassette deck. That distortion can be in a number of different forms, but the end result is usually the same: "fuzzy" sound.

Distortion also is contributed by the speakers, no matter how good the design. The goal is to minimize the distortion. The four basic types of speaker distortion are harmonic distortion, noise, transient response, and clipping.

Harmonic and Intermodulation Distortion

Harmonic and intermodulation distortion are characterized by the presence of frequencies in the speaker output which are not in the electrical input. These spurious signals are caused by imperfect driver and suspension behavior and are especially prominent when the speaker is driven to very high volume levels. Proper design of the speaker enclosure can significantly reduce this type of distortion.

Noise

Speaker noise is usually a raspy sound caused by a damaged component, such as a torn cone. Bits of acoustic batting, used to muffle the sound inside a speaker cabinet, is another common source of noise. The batting falls on the cone, and vibrates as the cone moves.

Transient Response

The transient response is the time delay as the speaker cone begins from rest to respond to a sudden, sharp, electrical pulse. The transient response (how fast the speaker reacts to make sound) depends on a number of factors, including the stiffness of the cone and suspension, and the enclosure design. A stiff cone and suspension act to resist movement, so the speaker may respond slower and cause a delay in the reproduced sound.

Clipping

Clipping occurs when the speaker cone cannot move as far as required by the audio signal. When the cone "bottoms out," the speaker is not able to accurately reproduce the sound. In fact, damage may result if the speaker is left operating in this condition, because the cone, bobbin, or suspension may eventually be torn. Clipping is most prevalent in woofers, where the cone must move a great distance to produce a high volume level at very low frequencies.

TYPES OF SPEAKERS

Speakers can be broken down into four major categories: woofer, midrange, tweeter, and full-range. Note that the frequency range overlaps to some extent across speaker types.

Woofer

The woofer reproduces the very low sounds. Their most efficient range is 20 Hz to 1000 Hz range, but they can extend to 3 kHz. As shown in *Figure 2-3*, woofers can be any size from about four inches in diameter to over 15 inches in diameter. A diameter of 10 to 12 inches is common in home hi-fi speaker systems. Felted paper and polypropylene are common woofer cone materials. Note in particular the 12-inch subwoofer. This speaker has dual voice coils and can provide 120 watts of bass range stereo power from the speaker.

Midrange

The midrange reproduces the broad spectrum of sound from about 1000 Hz to 10 kHz. Their most efficient range is 1 kHz to 4 kHz. The midrange speaker is responsible for playing back singing voices and most all instruments, yet it need not be physically large to do the job well. Midrange speakers come in sizes from about three inches to eight inches in diameter, as shown in *Figure 2-4*, with the smaller sizes more common.

4" LONG-THROW

6½" POLYPROPYLENE

12" POLYPROPYLENE

15" POLYPROPYLENE

12" SUBWOOFER
(DUAL VOICE COIL)

Figure 2-3. A selection of popular woofer speakers. *(Courtesy of Radio Shack)*

WIDE DISPERSION
SOFT DOME

3" MIDRANGE/TWEETER

Figure 2-4. A selection of popular midrange speakers. Note that the three-inch speaker shown is a midrange/tweeter combination. It can be used with a woofer in a two-speaker enclosure. *(Courtesy of Radio Shack)*

Most midrange speakers use paper, cloth, or polypropylene cones to reproduce sound. Some midrange speakers use a small plastic dome. The design of the dome provides a wider sound dispersion.

Tweeter

The tweeter reproduces the high or treble sounds in the range of about 4 kHz to 20 kHz (and beyond). Tweeters are usually small — under two inches in diameter. As shown in *Figure 2-5,* tweeters use a paper or cloth cone, a plastic or metal dome, or piezoelectric element and diaphragm to make sound.

Tweeters suffer the most from narrow sound dispersion — the signal travels a narrow corridor through the air. For maximum listening pleasure, the sound should be dispersed to cover a wider listening area. Tweeters are often outfitted with horns, baffles, and mechanical "lenses" to help disperse the sound. As with dome midrange speakers, dome tweeters have a wider dispersion than conventional cone designs.

1" MYLAR DOME

4" DYNAMIC CONE

WIDE DISPERSION
DUAL RADIAL HORN

DYNAMIC HORN

RECTANGULAR
PIEZO

Figure 2-5. A selection of tweeters. Various techniques are used to widen the dispersion of sound from tweeters, including domes and horns. Tweeters can be either electrodynamic or piezoelectric. *(Courtesy of Radio Shack)*

Full-Range

A full-range speaker is one that is engineered to adequately reproduce most of the audible sound spectrum. Full-range speakers usually represent a compromise over a system using individual woofer, midrange, and tweeter speakers, and their frequency response is not as good. No single speaker can accurately reproduce the entire range of human hearing. Full-range speakers are typically used in compact, inexpensive "bookshelf" enclosures.

Coaxial and Triaxial

A coaxial speaker combines a woofer and midrange, or midrange and tweeter, as one unit. A triaxial speaker combines a woofer, midrange, and tweeter as one unit. In each type, the speaker employs multiple cones and voice coils, surrounding a common heavy-duty magnet. Coaxial and triaxial speakers provide a marked improvement in frequency response over full-range speakers, yet take up little room. Many automobile manufacturers claim to have coaxial or triaxial speakers in their sound systems; however, they are not true coaxial or triaxial speakers because the speakers actually have individual magnets and voice coils for each cone.

Whizzer Cones

Some woofer and midrange speakers have a separate whizzer cone that acts like a tweeter. The whizzer is attached to the cone and voice coil of the speaker; unlike the coaxial and triaxial speakers described above, it lacks its own voice coil. The whizzer, shown in *Figure 2-6*, improves the high-frequency response of the speaker, but only marginally. The whizzer is placed around the dust cap in the center of the speaker cone.

SPEAKER DESIGN FACTORS

A speaker is a speaker, right? No way. Engineering didn't stop when the first speakers were invented. A variety of designs and materials are used in today's speakers, and each has its advantages over the competition. But it takes understanding of many factors that affect the performance. Let's look at several of these.

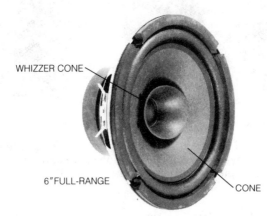

Figure 2-6. Whizzer cones are used on some full-range and woofer speakers to provide extended high frequency response. *(Courtesy of Radio Shack)*

Power Rating

Audio amplifiers are rated according to the amount of power they can deliver to the speakers. The power rating is in watts. Watts is a measure of the consumption of joules of energy per second. It is commonly expressed as the RMS (root-mean-square) voltage applied to an ac circuit times the RMS current in the circuit. In dc circuits, it is the dc voltage applied to the circuit times the current in the circuit.

You are probably most familiar with wattage as it relates to light bulbs. You know that high wattage bulbs put out more light than low wattage ones. The amplifier and speaker power rating is the same type of rating. It is the power delivered to the speakers due to the output signal voltage causing current through the voice coil.

Loudspeakers (especially woofers) often carry both a peak and an average wattage rating. The peak power rating is the maximum power the speaker can take for a very short time. The average power rating is the amount of power the speaker can accept continuously without damage. If the peak power rating is exceeded for more than a few seconds, there is danger of overheating and deforming the voice coil, which can cause permanent damage to the speaker. A typical wattage specification for a 10-inch to 12-inch woofer is 50 to 60 watts. Because midrange and tweeter cone speakers are smaller, and can't easily dissipate excess heat caused by extreme wattages, they have maximum power ratings of only 10 to 40 watts.

Some speakers have special cooling devices which dissipate heat quickly, even though the speaker itself may be physically small. Many high-end speakers, particularly tweeters, use a magnetic fluid called ferrofluid that keeps the driver elements cool. The cooling allows the speaker to handle higher power levels without burn-out or distortion. The ferrofluid also acts as a suspension for the inner circumference of the cone. The fluid replaces the mechanical spider, which in the typical speaker design, holds the inside of the cone to the frame.

Note that most amplifiers (and receivers) are rated in watts per channel. A 100 watt per channel amplifier delivers up to 100 watts RMS of reasonably distortionless sound to a set of speakers. The set of speakers — woofer, midrange, and tweeter (or whatever) — must have a combined power handling ability of at least 100 watts, or overheating and damage could result. Most of the power will go to the woofer. For symphonic music and most popular music, more than half the power lies in the three octaves between 75 and 500 Hz. Speakers are sometimes rated by their minimum power level. The recommended minimum power level is the power required to produce acceptable volume levels, say 85 to 95 dB SPL, in the average home listening room with a volume of 1500 to 1800 cubic feet.

Voice Coil and Magnet Size

One of the factors determining the power rating of a speaker is the size of its voice coil. A speaker with a high power rating uses a large voice coil which allows more heat to be dissipated; therefore, more power can be applied to the speaker. The diameter of the voice coil is often given, especially for woofers, but this figure does not relate directly to power rating. Therefore, you must rely instead on the maximum input power specification when choosing the right speaker for your application.

The weight of the magnet used in the speaker is almost always provided in speaker specification literature. Woofers use the largest magnets, because the speaker cones must be moved large distances to produce the low-frequency, high-volume sounds. A woofer with a magnet weighing 15 to 20 ounces is not uncommon. Larger, more capable woofers have even heavier magnets — up to 35 to 50 ounces.

The magnet material also is an important consideration. Magnets made of rare earth metals have more magnetic force than iron-nickel or ceramic magnets, thus, the rare earth magnet does not need to be as large to provide the same amount of magnetic force. Rare earth magnets are found mostly in tweeters and some midrange units. Keep magnet material in mind when comparing speakers.

Be wary of judging the quality of the speaker solely on the weight of the magnet. The larger the magnet, the greater the damping. An overdamped speaker will have reduced low-frequency response, but an increased high frequency response; therefore, the magnet size is a compromise. It must be large enough to provide the proper frequency response, but an excessively large magnet is overkill.

Cone Material

In the old days, speaker cones were invariably made of treated cotton or other natural fiber cloth. Now, speaker cones can be made of cloth, paper, a combination cloth and paper, or polypropylene plastic. Cones made with cloth and paper are subject to greater deterioration with age; polypropylene lasts longer and is able to take more stress when the speaker is operating at high volume. When cloth and paper are used, the exact make-up of the cone material is usually not provided in speaker literature. If polypropylene is used, it is usually specified.

Suspension

The suspension (or surround) largely determines the compliance of the speaker. With low compliance, the speaker is said to be stiff, and the cone will be difficult to move when you press it gently with your hand. With high compliance, the speaker is said to be flexible, and the cone will move easily when similarly pressed.

Suspensions are usually made with folded paper, rolled rubber, or rolled polyfoam, as illustrated in *Figure 2-7*. Folded paper suspensions provide a large amount of stiffness and are often used in ported reflex type speaker enclosures. Rolled polyfoam or butyl rubber suspensions permit the cone to move more freely, so speakers that use this design are often called "high compliance," and are the best suited to the acoustic suspension (sealed box) speaker enclosure. The type and design of suspension is of most importance for woofers and is usually given in the specifications. The specifications sheet that accompanies most midrange and tweeter speakers generally do not list the suspension type or material.

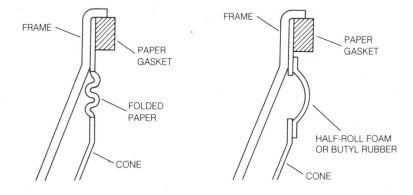

Figure 2-7. The design of the surround (or suspension) of woofer speakers varies. Woofers intended for ported reflex enclosures occasionally use stiff folded paper suspensions. The half-roll foam or rubber suspensions are not a stiff, and can be used in acoustic suspension (sealed box) enclosures.

Impedance

Impedance is a measurement of electrical resistance to ac signals, and is expressed in ohms. Most hi-fi speakers are rated at four or eight ohms. Sound quality is not enhanced or diminished by the choice of a higher or lower impedance. Rather, you choose four-ohm or eight-ohm speakers depending on the design of the system and how you connect the speakers together in a multiple-speaker enclosure. Details on wiring speakers is provided in Chapter 6. Most speakers are rated at eight ohms, and it is these you will most often use.

Note that the impedance of the speaker is different than its dc resistance, as measured at the speaker terminals. Measuring the dc resistance of an eight-ohm speaker with an ohmmeter usually yields a reading of five to six ohms.

Cone Resonance

Resonance is the tendency of a material to vibrate easily at a particular frequency. The singer uses resonance to break a crystal glass. By singing the particular note at which the glass material is resonant, the crystal can be made to vibrate enough so that it shatters. Bridges that are otherwise expertly built have been known to crumble to the ground when subjected to certain velocities of wind. The wind causes the bridge to swing back and forth, and if the period of oscillation matches the resonance of the bridge, dastardly things can happen.

The cone in hi-fi speakers have a resonance as well. The resonance is determined by the mass of the speaker cone, other moving parts (called the moving mass), and the suspension compliance. At a particular frequency, usually below 100 Hz for woofers, the cone will start to vibrate excessively when an electrical signal is applied, and the speaker will operate out of control. Not only is it unable to properly reproduce low frequency sounds, but it also colors (adds frequencies to) the sound output. Generally, the cone resonant frequency of a speaker sets a lower limit on its frequency response. Below resonance, acoustic output from the speaker falls off rapidly.

When the frequency of resonance is tested while the speaker is out in the open, the specification is termed free-air resonance, and is usually stated as a single frequency or a range of frequencies (such as 50 Hz ±10 Hz). Generally, the larger the speaker, the greater the mass and compliance, so the lower the free-air resonance. Speakers with heavy magnets also tend to have a lower free-air resonance. As you will see in Chapter 3, the resonance of a speaker is increased when it is placed in an enclosure. You will use the free-air resonance of the speaker together with other parameters to determine the optimum enclosure size for the woofer speaker you are using.

HOW TO READ SPEAKER SPECIFICATIONS

One of the most demanding tasks facing you as a speaker designer and builder is analyzing the specifications given with a particular speaker and knowing how the specifications will affect the sound of your assembled speaker system. You've already learned what most of the specifications mean; the following tells you how to interpret the specifications and how they relate to building your own speaker systems. Note that these specifications are not given for every speaker type, nor do all speaker manufacturers list them.

Impedance

Speaker impedance is usually either four or eight ohms. Choose the impedance that matches the type of wiring and crossover network you want to use (see Chapter 6).

Impedance is given for all speaker types.

Frequency Response

The wider the frequency response the better. Balance the frequency range with the flatness of the response. Most specification sheets include a frequency response graph, showing the ability to deliver the same level of loudness over a wide sonic spectrum. Other specification sheets may just give a frequency range and response tolerance like "100 Hz to 5 kHz ±3 dB." The frequency response graph for most speakers looks jagged — almost like a hilly mountain. This is normal. However, the flatter the top of the mountain, the better.

Frequency response is given for all speaker types.

Free-Air Resonance

Free-air resonance is the most common way to express cone resonance. The frequency is often given as a range, but for computational purposes, you may choose the figure in the middle of the range. You will use the free-air resonance figure to compute the best size of enclosure for the particular speaker you are using.

Free-air resonance is typically given for woofers only.

Moving Mass

Moving mass is the effective mass of all moving parts of the speaker. It includes the mass of the cone, bobbin, voice coil and dust cap. It also includes that part of the mass of the suspension and spider which move with the cone. The moving mass together with the speaker compliance (or stiffness) determine the cone resonant frequency. Moving mass (sometimes expressed as Mass or M) is given for woofers and some midrange speakers.

Q(ts)

The Q of a speaker denotes its resonance magnification, which takes into account the degree of damping of a speaker, and the tendency of the speaker to reach its maximum sound output level (peak) when operating at the free-air resonance frequency. Speakers are rated by their Q when out of an enclosure — the Q and the frequency of resonance are increased when the speaker is placed in a sealed box. The free-air Q of a speaker can range from about 0.2 to 1.5, with 0.4 or 0.5 being common. You use the Q(ts) specification when custom designing your own speaker enclosures.

Q is usually reported only for woofers.

V(as)

It is difficult to measure the compliance of a speaker other than to say that it is "high" or "low." The V(as) figure is a way to show speaker compliance by comparison. V(as) is the volume of air (in cubic feet or liters) which has the same compliance as the speaker's suspension. The larger the speaker, the larger the V(as). You use the V(as) specification when custom designing your own speaker enclosures.

V(as) is given only for woofers.

SPL (Sensitivity)

SPL stands for sound pressure level, and indicates the volume of sound produced by the speaker when it is fed one watt of electrical power in its operating frequency range. The measurement is typically made with a microphone placed one meter from the speaker (the specification will usually indicate if another test criterion has been used). The higher the SPL, the more volume the speaker generates for a given input power. A high value indicates an efficient speaker. A typical SPL for an eight-inch woofer is about 88 to 90 db. Larger 12-inch to 15-inch woofers may have an SPL rating of 91 to 93 dB.

SPL or equivalently, sensitivity, is given for all speakers.

Power Rating

The maximum number of watts RMS that can be safely fed to a speaker is typically indicated in the specifications as power rating, input power, or power handling capacity. If two power rating figures are used, one is the nominal rating (the amount of power that can be continuously applied to the speaker under normal circumstances), and the maximum rating (the maximum amount of power that can be applied for short periods of time without causing damage).

The maximum power rating is given for all speaker types. The higher the rating, the better.

Magnet Weight

As we stated previously, the weight of the magnet, without the mounting hardware or casing, is important when considering the efficiency and damping of a particular speaker. Other tests and specifications cover the characteristics contributed to by magnet weight. You do not use magnet weight when computing speaker enclosure size.

Magnet weight is given for woofers, and sometimes for midrange and full-range speakers. Most woofers have magnets that weigh from 15 to 20 ounces, but high-power woofers can have magnets as large as 50 ounces.

OFF TO A GOOD START

The speakers themselves are but one element in speaker systems. The best speakers will sound "muddy" and "fuzzy" if improperly installed in an enclosure, and your choice of enclosure size and type is as critical as your choice of speaker. In the next chapter, you'll learn about speaker enclosures and how they are built, and how the the various speaker types are used to obtain optimum sound quality.

A CLOSE LOOK
AT SPEAKER
ENCLOSURES

Hi-fi speakers aren't complete without enclosures. The enclosure serves a variety of functions:

- It houses the speakers.
- It protects the speakers against damage.
- It improves the frequency response so that the sound is more natural.

Simply mounting a speaker in a box isn't enough; the enclosure must be designed so that it complements the speakers. The enclosure must be built to proper proportions and dimensions, or the sound it reproduces may be too boomy or too flat.

This chapter details the design specifications for high fidelity speaker systems. You'll learn why speakers benefit from installation in an enclosure, how the design of the enclosure determines the richness of sound, and how to calculate the proper proportions and size of an enclosure for your speakers.

WHY USE ENCLOSURES

As you learned in Chapter 1, sound waves are created by vibrations of the speaker cone. In open air, sound waves are dispersed in all directions from the speaker.

When the cone moves outward, it creates a positive pressure on the air in front of it and simultaneously creates a negative pressure (partial vacuum) on the air behind it. At low frequencies, where the cone diameter is much shorter than a wavelength, out of phase waves from the rear of the cone mix with and cancel the front wave, greatly reducing speaker output.

When placed in a suitable enclosure, the sound waves emitted from the rear of the speaker can't travel to the front of the cone. This prevents low frequency cancellation. As a result, speaker efficiency and sound output in the bass frequencies are greatly improved.

TYPES OF ENCLOSURES

A number of speaker enclosure designs have been devised in the past several decades, but the most popular are acoustic suspension and ported reflex. Both types are in wide use in off-the-shelf stereo systems, and both have their advantages and disadvantages.

Acoustic Suspension Enclosure

In the acoustic suspension enclosure, the box is sealed and is effectively airtight. The movement of the speaker acts as a piston to compress and decompress the air, as illustrated in *Figure 3-1*. Acoustic suspension designs decrease the compliance of the speaker, making it stiff and increasing its resonant frequency and Q relative to the free-air values. This improves bass response, but sacrifices some efficiency relative to ported reflex designs. The amplifier has to deliver more power to produce the same volume.

Ported Reflex Enclosure

In the ported reflex design, a small hole (usually two to three inches in diameter) is cut into the enclosure. In most designs, a tube is then inserted in the hole. As shown in *Figure 3-2*, the tube acts as an air duct, providing a partial vent for the compressing and decompressing air. As stated previously, the ducted port together with the enclosed air volume form what is called a Helmholtz resonator. The action of the resonator is very much like the effect one gets when blowing over the mouth of a large bottle or jug. Speaker cone motion excites the resonator and it emits sound from the port. When the port/box combination is properly tuned, low frequency air vibrations within the port are in phase with cone motion and output is increased. The deep bass efficiency of a ported reflex design can be 50 to 100% greater than that of an acoustic suspension system. The cross-sectional area and length of the duct together with the box volume determine the tuned frequency of the enclosure.

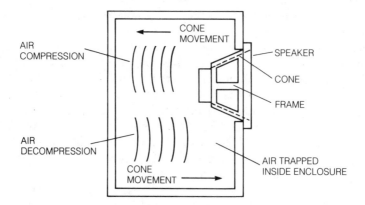

Figure 3-1. The cone of the speaker acts as a piston, compressing and decompressing the trapped air in the enclosure.

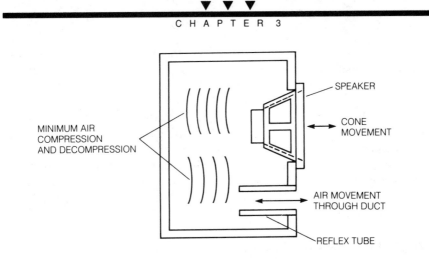

Figure 3-2. The duct provides a passageway that lowers the pressure inside the enclosure during speaker cone movement. The effect is an increased effective enclosure volume.

Acoustic suspension enclosures are the easiest to design and build. Only rudimentary calculations based on the speakers you are using are required to determine the best dimensions for the enclosure. Ported reflex enclosures are slightly harder to build because you must also calculate the diameter and length of the duct based on the internal volume of the enclosure and the characteristics of the woofer.

You may wish to build one or two acoustic suspension designs, then with your hands-on experience, you can graduate to the more sophisticated ported reflex enclosures. Although ported reflex enclosures are more demanding, they are worth the effort because they generally offer higher efficiency and deeper bass response.

ACOUSTIC SUSPENSION ENCLOSURE DESIGN

Acoustic suspension enclosures can be built to almost any size. For small diameter speakers, the enclosure can be just slightly larger than a few thick paperback novels. Because you can make them so small, acoustic suspension enclosures are ideal for "bookshelf" designs. These compact speaker enclosures can fit on a bookshelf beside a small stereo system, or on the fireplace mantle.

Don't let size stop you, however, because you can construct large acoustic suspension enclosures just as easily as small ones. A three-way enclosure with a 12-inch woofer, for example, might measure about 30 by 17½ by 11 inches (HWD). Because they are quite large, enclosures of this size are usually too large for a bookshelf and must be placed on speaker stands.

Choosing the Speakers

High compliance woofers are best suited for small acoustic suspension enclosures. That means that the surround on the woofer is fairly flexible. When placed in a relatively small sealed enclosure, the air trapped in the box acts like a spring to stiffen the cone. This lowers the speaker compliance and increases its resonant frequency above the free-air value. You may use a low compliance woofer if the enclosure is physically large. The large volume of the enclosure makes up for the low compliance of the speaker.

Recall from Chapter 2 that the compliance of woofers is specified as V(as), which is the equivalent volume of air that has the same compliance as the speaker. A few low-cost woofers do not list the V(as), but rather state the compliance as "high" or "low." If the speaker compliance is not stated, or is implied, you can estimate the compliance by inspecting the type of surround used. Because of their "bounciness," polyfoam and rubber roll surrounds offer high compliance. Rolled or folded paper surrounds are stiffer, which lowers the compliance of the speaker.

Estimating Enclosure Volume

The volume of the enclosure is determined by the size and characteristics of the woofer. The midrange and tweeter are not a major consideration when estimating enclosure volume, although you must take them into account to allow space for them in the box. Elaborate design equations, given in Appendix A, can be used to calculate enclosure volume, and you may use these equations for your own custom-designed speaker enclosures. The equations, which take into account such woofer specifications as free-air resonance, Q(ts) and V(as), must be used when designing enclosures using speakers with unusual characteristics, such as high V(as) or high Q(ts).

An easier way to estimate enclosure volume is to use *Figure 3-3*. The illustration shows acceptable enclosure volumes for woofers of varying sizes from 4 inches to 15 inches. The calculations for enclosure volume are already done for you, based on woofers of average design and average characteristics.

Note that there is a great deal of flexibility in enclosure volume. A typical 12-inch woofer, for example, may be housed in an enclosure that has a volume of between 1.75 and 3.5 cubic feet. This wide variance makes it easy to match individual speaker specifications to a universal size enclosure. You can use the chart in *Figure 3-3* as the basis for your first custom-designed speaker enclosures. Then, when you gain practical knowledge of building your own enclosures, you can use the formulas in Appendix A to fine-tune the enclosure volume to match your specific woofer.

You must use the equations for a speaker like the Radio Shack 12-inch dual-voice-coil subwoofer (see *Figure 2-3*) which has a high V(as) of 13.3 cubic feet, a Q(ts) of 0.380, and a free-air resonance frequency of 21 Hz. The enclosure volume is 2.5 cubic feet when Qb = 0.95.

Measurement Conversions

You'll be working with feet and inches when building speaker enclosures and you'll want a way to convert linear inches into cubic inches. You'll also want an easy way to convert cubic feet to cubic inches, and vice versa. Here's how to do the math:

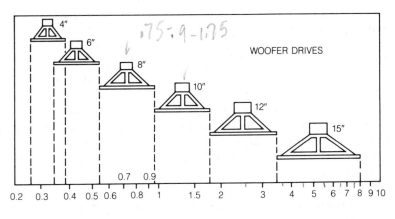

Enclosure Volume — ft³ (Cubic Feet)

Figure 3-3. The size of the woofer is the single most important factor that determines proper volume of the speaker enclosure. Use this chart, which is based on typical woofers, to estimate enclosure volume.

- To convert linear inches to cubic inches, multiply the height in inches times the width in inches times the depth in inches. (Example: 16 inches times 12 inches times 8 inches = 1536 cubic inches.) A linear foot is 12 inches; therefore, a cubic foot is 12 inches times 12 inches times 12 inches = 1728 cubic inches.

- To convert cubic inches to cubic feet, divide the number of cubic inches by 12 three times. (Example: 1536 cubic inches divided by 12 divided by 12 divided by 12 = 0.89 cubic feet.) Since 1728 cubic inches equals one cubic foot, another way to do it is to divide the number of cubic inches by 1728 to obtain cubic feet. (Example: 1536 cubic inches divided by 1728 = 0.89 cubic feet.)

- To convert cubic feet to cubic inches, multiply the number of cubic feet by 12 three times. (Example: 0.89 cubic feet times 12 times 12 times 12 = 1538 cubic inches. Since one cubic foot equals 1728 cubic inches, another way to do it is to multiply the number of cubic feet by 1728 to obtain cubic inches. (Example: 0.89 cubic feet times 1728 = 1538 cubic inches.)

Calculating Enclosure Size

Speaker designers have long used the "golden ratio" when calculating the dimensions and proportions of speaker enclosures. The ratio used is 0.6:1:1.6 (depth, width, height, respectively). For example, if the enclosure is one foot wide, the remaining golden ratio dimensions are 0.6 feet deep and 1.6 feet high.

In practice, the golden ratio can't always be exactly followed, but you should strive for the ideal proportions whenever possible. The rectangular design enhances sound quality. At the very least, avoid square-box enclosure designs because they tend to exaggerate some frequencies over others.

The projects in Chapter 8 present ready-made plans for speaker enclosures, but if you want to design your own enclosures, use the graph in *Figure 3-4* as a guide. It shows enclosure dimensions based on the 0.6:1:1.6 golden ratio.

To use the graph, follow these three steps:

1. Use *Figure 3-3* or compute the approximate required volume based on the woofer you are using.
2. Find the corresponding cubic feet value on the horizontal scale.
3. Read the dimensions for the depth, width, and height at the intersections of the vertical line with each of the curved lines.

Let's say, for instance, that you have calculated the required volume of an enclosure based on the design parameters discussed in the previous section. You are using a 12-inch woofer and have settled on an enclosure volume of 2.5 cubic feet. The graph in *Figure 3-4* indicates that for this enclosure, the ideal dimensions are:

- Depth: 10 inches
- Width: 16¾ inches
- Height: 26½ inches

Remember these are inside dimensions of the enclosure so measure you speaker layout accordingly.

Maximizing the Speaker Layout

Before you cut the wood for the enclosure using these dimensions, you should double check to make sure that the size of the box matches the speakers you are using. You also need to make sure that the enclosure will fit the allocated space in your home. Should the box be too small to house the components, or too large for the space in which you want to put it, increase or decrease the enclosure volume until the dimensions are correct.

Arranging Speakers

To calculate the area needed to adequately house the speakers, lay each speaker on a flat surface and arrange them in a straight line. Organize the speakers so that the woofer is at one end, the midrange (if used) is in the middle, and the tweeter is at the other end. Allow for at least 1/2-inch clearance between speakers. At no time should the speaker frames touch. If your speaker design uses L-pads, remember to leave room for them.

Measuring Your Speakers

Measure the distance from the bottom of the woofer to the top of the tweeter and add two inches to provide space between the speaker and the top and bottom panels of the enclosure. This is the minimum height (H) dimension for the enclosure. Measure the depth of the woofer (from front to back) and the width of the woofer and add one inch to each measurement. These are the minimum depth (D) and width (W) dimensions, respectively, for the enclosure. All of these dimensions are inside dimensions.

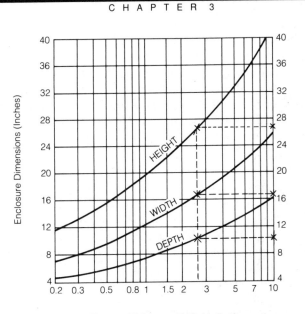

Enclosure Volume — ft³ (Cubic Feet)

Figure 3-4. Use this graph to compute the ideal dimensions for speaker enclosures of a given volume. Find the desired volume on the horizontal scale and read the height, width, and depth at the intersection of the curve lines.

Your measurements should be less than the dimensions of the enclosure obtained from the chart in *Figure 3-4*. If the height measurement is much less than then height of the enclosure, you may decrease the size of the enclosure only if the volume remains within the acceptable design limits of the woofer you are using. If any of the measurements are more, you need to either increase the size of the enclosure, rearrange the speakers, or choose a smaller midrange and/or tweeter.

Increasing Enclosure Size

Increasing the size of the enclosure should be done only if you do not exceed the volume limit for the woofer you are using. For example, if the woofer calls for a volume of between 1 and 1.75 cubic feet, you should not increase the size of the enclosure so that the volume exceeds 1.75 cubic feet.

The overall sound quality of the enclosure generally is not affected by changes in dimensions of less than 10 percent. However, rather than increase all the dimensions by a few inches to accommodate large speakers, you may safely increase only one of the dimensions — the height, width, or depth — as long as the change is under 10 percent.

For example, you may safely increase the 30-inch height dimension of an enclosure by three inches (10 percent of 30 inches) to 33 inches without adversely affecting the sound quality of the speaker system. Just to be sure, compute the new volume to see if the enclosure still conforms to design specifications.

Rearranging Speakers

You must exercise care when rearranging the speakers. The best results are obtained when the speakers are arranged in a straight vertical line. You may, if absolutely necessary, position the tweeter and midrange (if used) slightly off the center axis of the speaker line. Repeat the off-axis alignment, but in mirror image, for the right-hand speaker enclosure. Never use an arrangement where the midrange and tweeter are on the same horizontal axis.

Choosing Smaller Speakers

When you can't change the size of the enclosure, and positioning the speakers off-axis doesn't solve the problem, your remaining alternative is to use smaller tweeters and midrange speakers. These speakers are not as critical as the woofer and you have greater flexibility in changing them. You might, for instance, exchange a medium-size dome tweeter for a smaller piezoelectric horn tweeter. Note that cramped enclosure designs are best suited for two-way speaker systems that use just a woofer and tweeter.

PORTED REFLEX ENCLOSURE DESIGN

The port (hole) and duct (tube) in a ported reflex speaker let you use relatively small enclosures and still obtain deep bass response. When the box is properly tuned, the air vibrating within the port is in phase with the very low frequencies of the woofer. This accentuates the bass of the speaker, making it sound richer.

The basic design of ported reflex speaker systems is similar to acoustic suspension. In fact, you calculate the depth, width, and height in the same way as for the acoustic suspension enclosure. The difference is that you have to compute the proper opening and length of the port. As in a clarinet, saxophone, or other wind instrument, the length of the duct and the area of the opening determine the operating frequency. For a fixed length duct, increasing its diameter will raise the reflex enclosure frequency. Conversely, decreasing the diameter will lower the frequency. For a given diameter, increasing the length of the duct will lower the enclosure frequency, while shortening the duct will raise it.

Port Tuning Characteristics

The opening for most ported reflex speakers varies between two and three inches in diameter, although it can be larger. Duct length varies from about one inch to 10 inches or more. The duct for ported reflex speakers is usually placed in the speaker mounting board. However, unusual designs may place the duct on the bottom, side, or even back of the enclosure. In this book, we'll concentrate on conventional ported reflex enclosure designs where the duct is mounted in the speaker mounting board. Before you calculate the port dimen-

sions, you must consult the specifications sheet that came with the woofer and find the following specifications:
- Free air resonance (fo or fs)
- Q(ts)
- V(as)

Let's say that the 10-inch woofer you are using has the following specifications (these are fairly typical for a 10- to 12-inch woofer):
- Free air resonance, fs = 40 Hz
- Q = 0.53 (same as Qts)
- V(as) = 2.67 cubic feet.

Some speaker manufacturers list V(as) in liters. Divide liters by 28.32 to obtain cubic feet. For example, a V(as) of 160 liters is equal to 5.65 cubic feet.

Calculating Port Characteristics

With the help of some graphs, only simple arithmetic is required to calculate the port characteristics. These are for *typical* woofers only. Use the equations in Appendix A if speaker has unusual characteristics such as high V(as) or Q(ts). The calculations are done in three steps:

1. Calculate the optimum volume of the enclosure.
2. Calculate the tuning frequency for the enclosure.
3. Calculate the duct size and length.

Calculating Optimum Volume

Use the graph in *Figure 3-5* to calculate the optimum volume of the enclosure. Find the value in the horizontal scale that matches the Q of the woofer, in this case 0.53. Extend a line from this point upward until you intersect the curved line, then read the corresponding volume multiplier on the vertical axis. According to the chart, the volume multiplier for a speaker with a Q of 0.53 is about 2.5

To complete this portion of the calculation, multiply this volume multiplier by the speaker's V(as). For the example, the calculation is:

$$2.5 \times 2.67 = 6.75 \text{ cubic feet}$$

Calculating Tuning Frequency

Use the graph in *Figure 3-6* to calculate the tuning frequency factor for the enclosure. Find the value in the horizontal scale that matches the Q of the woofer (0.53). Extend a line from this point upward until it intersects the curved line, then read the corresponding tuning frequency factor on the vertical axis. According to the chart, the tuning frequency factor for a speaker with a Q of 0.53 is about 0.72.

Next, multiply 0.72 by the free-air resonance of the speaker, in this case 40 Hz. The result is the tuning frequency of the enclosure. For the example, the calculation is:

$$0.72 \times 40 = 28.8 \text{ Hz}$$

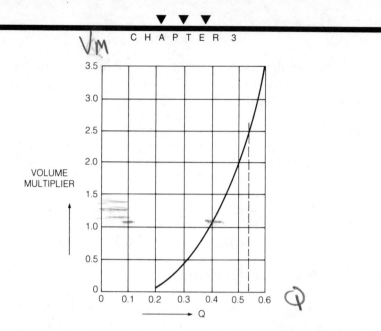

Figure 3-5. Use this graph to determine the ideal volume for ported reflex enclosures.

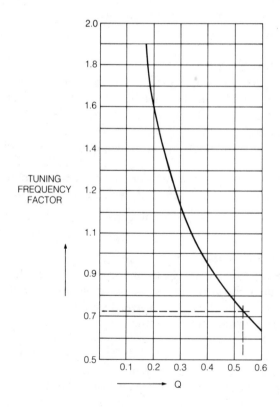

Figure 3-6. Use this graph to determine the ideal tuning for the duct in ported reflex enclosures.

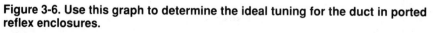

Calculating Duct Size and Length

Use *Tables 3-1* and *3-2* to calculate the size and length of the duct for the enclosure. *Table 3-1* shows the duct lengths for a 2-inch inside diameter tube and *Table 3-2* shows the duct lengths for a 3-inch inside diameter tube. For larger speakers, try to make your design work with the larger diameter duct as this will reduce air particle velocity in the port and the possibility of port noise.

Using the volume of 6.75 cubic feet and tuning frequency of 28.8 Hz, *Table 3-2* (3-inch tube) shows that a tube with a length of approximately 1.5 inches (about midway between the values in the 25 and 30 Hz columns) is ideal for this enclosure.

Table 3-1. Length of Two-Inch Tube in Inches*

	Frequency (Hz)											
Volume Cu. Ft.	**20**	**25**	**30**	**35**	**40**	**45**	**50**	**60**	**70**	**80**	**90**	**100**
0.50						7	5⅜	3¼	2	1¼		
0.75					5⅝	4	3	1⅝	⅞			
1.00				5½	3⅞	2¾	1⅞	⅞				
1.25			6	4	2¾	1⅞	1¼					
1.50		7⅝	4¾	3⅛	2	1⅜	¾					
1.75		6¼	4	2½	1½	1						
2.00		5⅜	3¼	2	1¼							
2.50	7	4	2¼	1¼								
3.00	5⅝	3	1⅝	⅞								
3.50	4⅝	2½	1¼									
4.00	3⅞	2	1									
5.00	2¾											

*Recommended duct lengths using two-inch inside diameter tubing.

Table 3-2. Length of Three-Inch Tube in Inches*

	Frequency (Hz)											
Volume Cu. Ft.	**20**	**25**	**30**	**35**	**40**	**45**	**50**	**60**	**70**	**80**	**90**	**100**
0.50								8⅜	5⅝	3¾	2½	1⅝
0.75						10¼	7⅞	4¾	3	1¾	⅞	
1.00					9⅝	7⅛	5⅜	3	1⅝	¾		
1.25					7¼	5¼	3⅞	2	⅞			
1.50				8⅛	5¾	4	2⅞	1¼				
1.75				6⅝	4½	3⅛	2⅛	¾				
2.00			8⅜	5½	3¾	2½	1⅝					
2.50			6¼	4	2½	1½	¾					
3.00		8	4¾	3	1¾	⅞						
3.50		6½	3¾	2¼	1¼							
4.00	9⅝	5⅜	3	1⅝	¾							
5.00	7¼	3⅞	2	⅞								
6.00	5¾	2⅞	1¼									
7.00	4½	2⅛	¾									
8.00	3¾	1⅝										
10	2¼	⅞										
12	1¾											

*Recommended duct lengths using three-inch inside diameter tubing.

Duct Construction

You can use plastic, paper, or metal for the duct. However, metal ducts may vibrate and produce noise, so plastic or paper are better choices. Plastic drain pipe, commonly available in both 2- and 3-inch sizes, is available at most plumbing and hardware stores. The drain pipe is a good choice when the tube must be longer than the depth of the enclosure. You can use elbow fittings to make an "L" inside the enclosure, and extend the length of pipe upward or sideways inside the box, as depicted in *Figure 3-7*. An alternate solution is to use a smaller diameter, shorter length duct.

A carpet store may sell (or give) you a heavy cardboard tube used as the core for a roll of carpeting material. Obtain a tube with a 2- or 3-inch internal diameter and use a hacksaw to cut it to size. Trim off any bits of paper from the ends of the tube to prevent an irritating flapping noise when air passes through.

Placement of the Duct

The duct can be placed just about anywhere on the front of the enclosure. Most speaker designers prefer to place the duct close to the woofer, but not closer than about three inches. A spot midway between the woofer and tweeter is often used.

Before cutting the hole for the duct, examine the mounting scheme for all the other components in the system. If you will be installing additional components, such as crossover networks and L-pads, you must allow room for them. Be sure that their placement and mounting do not interfere with the duct.

Before cutting the hole for the duct, measure the *outside* diameter of the tube with a ruler or caliper. Cut the hole exactly to size, or slightly smaller. By cutting the hole slightly smaller, the tube fits very tightly in the hole so that little gluing and caulking is required. Use a mallet to drive the tube into the

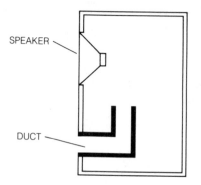

Figure 3-7. Bend the duct in an L-shape if the duct extends to within two to three inches of the back panel of the enclosure).

hole until it is flush with the front of the panel. If the hole is too small, file it lightly to enlarge it. Glue the tube in place using filling wood glue and apply caulking around the tube on the inside of the enclosure.

Experimenting with Box Tuning

Although the formulas detailed earlier provide the best means of calculating the proper length and opening for the duct, you may wish to experiment with the tuning of the enclosure by varying the length of the tube. Calculate the recommended tube length and, if possible, add a few more inches. Complete the enclosure, but mount the front panel in a temporary manner so you can easily remove it. Try the system for a while and carefully listen to its performance. You may wish to use a sound level meter and a test record or tape to obtain a more objective analysis of the speaker system's frequency response.

If the enclosure is tuned too high, bass response will be excessive and the speaker will sound boomy. If the enclosure is tuned too low, bass response will be weak. Since you made the duct too long, the box probably will be tuned too low, so the bass should sound bland. Increase the tuning frequency by cutting off a short piece of the tube.

THE ROLE OF DAMPING

Adding damping material, such as fiberglass batts, to the enclosure changes the effective volume of the box. Attach the damping material to the inside walls of the enclosure. As you add damping, the effective volume of the enclosure *increases*. This is due to complex acoustic conditions which are beyond the scope of this book.

You should add damping to all speaker enclosures; the graphs and formulas presented earlier in this chapter assume that you will add a one-inch thick layer of damping to the bottom, sides, and back of the enclosure. Don't put damping material on the top because it may come loose and fall onto the speakers. Without the damping, the sound from the speaker may sound artificially hollow.

You can also use damping to correct mistakes in enclosure design. Should you build a speaker box that turns out to be too small for the woofer you have chosen, you can add another layer or two of damping material to increase the effective volume. Although this technique is helpful in some situations, be aware that you can achieve no more than about a 10 to 20 percent increase in effective volume. Too much damping can reduce efficiency and lower bass output, especially in ported reflex designs.

ENCLOSURE CONSTRUCTION

The dimensions provided by the graphs and formulas presented in this chapter are for the interior of the enclosure. When cutting the wood for building the enclosure, you must add some to the dimensions to account for the thickness of the stock. Assume, for example, that you are building an enclosure using 1/2-inch plywood. An internal width of 12 inches means you must cut the top and bottom pieces to 13 inches wide (12 inches plus 1/2 inch for each side), as shown in *Figure 3-8.*

Building speaker enclosures is not an exact science, so you don't have to build an enclosure that *precisely* matches the design formula. If your calculations show a height of 32³⁄₁₆ inches, for instance, you may safely round down the measurement to the nearest half-inch, or even the nearest inch.

Keep the following tips in mind when designing and building your own speaker enclosures:

- Be sure that the enclosure is rattle-free. Secure all components, including tuning duct, speakers, terminals, and crossover networks to the enclosure.
- Staple or nail the damping material to the inside bottom, sides, and back of the enclosure. Avoid placing damping material on the inside top as it may come loose and fall onto the speakers.
- Keep the damping material away from the duct opening.
- Be sure that the enclosure is completely airtight (except for the duct in the ported designs). Caulk the inside joints of the enclosure.

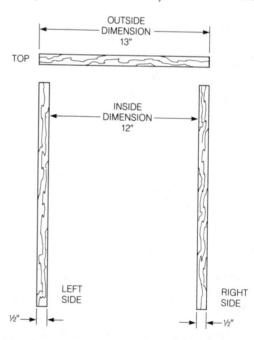

Figure 3-8. Calculations for enclosure dimensions are for the inside; add the appropriate amount to compensate for the thickness of the wood.

FINISHED WITH THE BASICS

With the information contained in Chapter 2 and this chapter, you are now ready to start designing and building your own speaker enclosures. In the next chapter, you'll learn about the materials used for speaker enclosures and how to assemble them using ordinary carpentry techniques.

CONSTRUCTION TECHNIQUES

Even the best speakers and the most careful planning will be for nothing if your speaker enclosures are made with the wrong materials. Choosing the wood, hardware, and other speaker enclosure parts requires the same diligence as shopping for the best woofer, midrange, and tweeter. All the materials you need to build quality speaker enclosures can be found at any good lumberyard, handyman store, or home improvement store. An otherwise good speaker design — even with the right materials — may sound muddy and harsh if the enclosure is not constructed with the proper techniques. In this chapter and Chapter 5, you'll learn how to construct professional-looking and professional-sounding speaker enclosures.

IDEAL WOODS FOR SPEAKER ENCLOSURES

There is an almost endless variety of woods, but only a few are truly suitable — and economically appropriate — for use in speaker enclosures. Particle board and plywood are especially well-suited to building quality speaker systems.

Particle Board

Particle board is sawdust and wood chips that have been glued together and pressed into a flat sheet. It routinely comes in 4 by 8 foot sheets, though you can often buy it already cut into smaller sheets. You can get a full 4 by 8 foot sheet for under $15. Choose a high density type.

Particle board is available in different thicknesses, from 1/4 inch to 1⅞ inch. A 5/8 or 3/4 inch thickness is ideal for speaker enclosures. The dense type of particle board makes it especially well-suited for use in speaker enclosures. Particle board can be effectively used for the parts of the enclosure that do not show, such as the grille board (a front piece with holes cut out), grille frame, enclosure back or bottom, and the actual board used to mount the speakers.

The disadvantage of particle board is that it is harder to work with than conventional solid woods, or even plywood. You can't easily shape particle board or make special joints. Chunks of particle board can flake off if the wood is mishandled, if dull tools are used to cut it, or if you use a nail that's too large or too long. In fact, it's much better to use predrilled holes and screws than any kind of nails for joining particle board.

What's more, particle board has no grain or natural luster, unlike even the most inexpensive woods. Thus, some sort of covering is required to keep the enclosures from being rather ugly. Fortunately, veneers and other laminates can be added to the outside to make very attractive enclosures. Details on finishing techniques are provided in Chapter 5.

Plywood

Plywood is another excellent wood for use in speaker enclosures. It is made by laminating several thin layers of wood together. The wood is oriented so that the grain changes direction at each layer. This gives plywood increased strength and durability.

Like particle board, plywood routinely comes in 4 by 8 foot sheets; smaller sheets are available at many lumber yards and home improvement stores. Thicknesses range from 1/4 inch to 1⅞ inch. The 3/4 inch thick plywood is a good choice for most speaker enclosures. Plywood comes in many grades. Grade A is the best, D the worst. Grade A means that there are no blemishes in the surface (such as plugged knotholes) and that there are no voids (empty spots) in the laminated layers underneath.

Both sides of the plywood are graded, so the typical sheet may be identified with something like A-C. That means one side is grade A, the other side is grade C. *Table 4-1* lists the common grades, their appearance, and their recommended use in speaker enclosures.

Since no one is interested in how your speaker enclosures look from the inside, you can use the less expensive A-B or A-C grade plywood. Construct the speaker enclosure so that the best side faces out. If you want a really good-looking enclosure, get a sheet of plywood with high-quality expensive veneer already applied to one side. The grade for the veneered side may be indicated by an "N."

Table 4-1. Plywood Grades

Grade (Front-Back)	Use in Speaker Enclosure Construction
A-A	Exterior pieces; veneered sides can be stained or oiled. High quality veneer on both sides is not important for most speaker enclosures, however.
A-B	Exterior pieces; use front (good side) for outside and back for inside.
A-C	Exterior pieces; use front (good side) for outside and back for inside. Also can be used for grille board, speaker mounting board, back, and battens.
C-D (shop grade)	Battens, bottom, and back. Not for any exposed exterior surface.

Hardwood Trim and Molding

Trim and molding can be added to enhance the appearance of your speaker enclosure projects. The trim or molding is usually applied around the front of the enclosure, or it can be used as the framework for the speaker grille. Trim and molding is available in many sizes, shapes, and varieties.

WORKING WITH WOOD

Only a few basic shop tools and supplies are needed to make impressive-looking speaker enclosures. At a minimum, you need:

- A power saw for cutting large pieces of lumber. A circular saw is a good, all-around choice. Use a guide fence whenever possible to ensure straight and square cuts. A radial arm saw or table saw should give better results because the wood is held firmly against square surfaces.
- A small backsaw or handsaw, for cutting small pieces of wood.
- A power saber saw, scroll saw, or jig saw for cutting openings for the speakers. Of course, you can use a manual coping saw if you prefer.
- A 16-ounce claw hammer.
- C-clamps and band-clamps, for holding glued pieces together as they dry.
- Measuring tape, folding rule, or yardstick.
- Carpenter's square.
- Assorted wood rasps, files, and sandpaper.
- Compass.
- Nail set.
- Screwdrivers (slotted and Phillips, various sizes).
- Drill and drill bits.
- Staple gun (not an office stapler) for attaching acoustic cloth and damping material.
- White glue or wood (carpenter's) glue. Wood glue is a similar mixture to white glue, but has extra ingredients to bond wood.

If you don't own a power saw, consider renting one for the duration of your speaker project. Alternatively, see if the folks at the lumber yard or home improvement store will cut the wood to size for you. Many offer this service for a nominal per-cut charge. You are nearly guaranteed straight, perfect cuts, and the smaller pieces are easier to transport home.

Layout and Cutting

Before you can cut the wood, however, you must figure out how the wood is to be cut to make all the pieces. Careful layout can save you time and money. One easy way to lay out the pieces for cutting is to prepare a worksheet using graph paper. Try to keep the grain of the wood in the same direction for all pieces. In most speaker enclosures, the grain is oriented vertically for the front and side pieces, and side-to-side for the top piece. *Figure 4-1* shows the right and wrong ways to lay out the pieces for cutting.

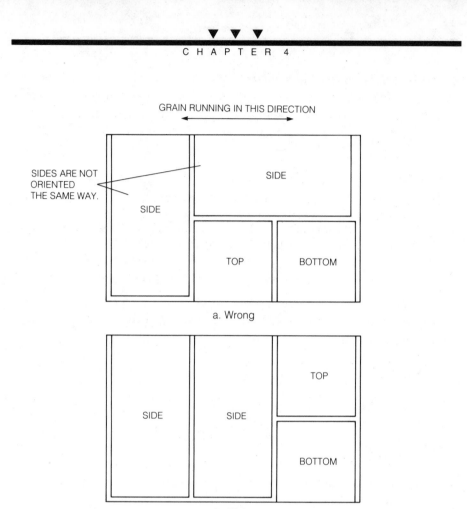

Figure 4-1. Pieces should always be oriented in the same direction so that the wood grain direction matches for the entire enclosure. You need not worry about grain direction when using particle board.

To ensure accuracy, use a calculator to total up the actual dimensions of the pieces to be sure they fit the area of the plywood or particle board. You may find it easier to convert all measurements to inches; for example, a 4 by 8 foot piece of plywood is 48 by 96 inches. After the layout is complete, you can mark the wood for cutting.

Remember that you lose a small amount for each cut — this is the kerf or width of the blade doing the cutting. As a result, you can't get four one-foot wide pieces across the four-foot dimension of the plywood.

The cutting procedure varies depending on the type of wood and the saw you are using. Consult any good carpentry book, available at a local library, for specific details. Use a cross-cut blade when cutting planks and particle board; a combination or plywood blade when cutting plywood. The resin composition of particle board dulls the cutting blade rather quickly, so be sure yours is new or recently sharpened. After many cuts, get the blade sharpened. *Always exercise caution when using power tools, and never defeat any safety feature. They are there to prevent injury. Be sure to wear safety goggles when operating any powered woodworking equipment.*

The veneer face of plywood is very thin, and can be damaged if you cut the wood carelessly. Keep the good face of the plywood up when using a handsaw, radial arm saw, or table saw, but turn it down when using a hand-held circular saw or motorized saber saw. This reduces splintering on the good side. You can further reduce splintering by backing the plywood with a piece of scrap and holding them together with clamps. Of course, this means you must cut through two pieces of wood instead of just one. You can also minimize splintering the veneer by using a sharp utility knife to make an incision approximately 1/16 inch deep along the cut lines before sawing. Use a carpenter's square to ensure a straight line and exercise extreme caution. Do not let the blade slip and cut your hands.

Cutouts for the speakers are made on the speaker mounting board and, optionally, on the grill board. You make these cuts with a motorized saber saw or scroll saw after drilling a starting hole, as shown in *Figure 4-2*. Use a blade recommended for the wood you are using.

To make the cutout, mark the placement of the speaker, following the guidelines in Chapter 3. Use a speaker to mark the outside dimension of the flange. Draw the outline directly on the wood. Next, measure the width of the mounting flange, and use this measurement to make a smaller, inner circle. This is the one you will cut out. Use a compass, string, or piece of wood to accurately draw the circle. Drill a hole inside the smaller circle near the circle line. The hole must be large enough for the saw blade to go through so you can cut out the circle.

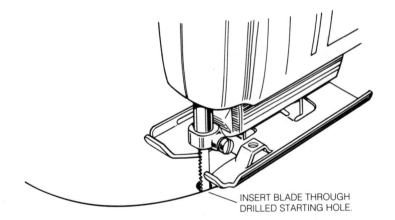

INSERT BLADE THROUGH
DRILLED STARTING HOLE.

Figure 4-2. Cutouts are best made with a power saber saw. (A model with variable speed provides more control over the cutting.) Drill a hole inside the cutout as a starting point for the blade.

Use the same technique to cut the holes for the ports in a bass reflex speaker. You can make holes for L-pads and fuses using a saber saw, but if the hole is small enough, use a spade drill bit. Make a "pilot hole" with a very small bit, then use the spade bit of the proper size.

Joint Types

You'll be making mostly straight cuts with your saw. The top and side joints of the speaker enclosure, however, may require a bit of carpentry knowledge.

The easiest way to join two pieces of wood is with a butt joint because it doesn't require any special tools or expertise. The butt joint is fairly strong, especially when reinforced with glue blocks as shown in *Figure 4-3a*. However, its finished appearance is not very desirable, especially for a speaker enclosure destined to be part of the living room decor. Unless you cover the enclosure with veneer or some kind of laminate, you can see the outlines of the lumber and how the pieces are joined.

A miter joint, as shown in *Figure 4-3b*, looks far more professional than the simple butt joint, and it's also a little stronger. You make miters by cutting the ends of the wood at a 45 degree angle, picture frame style. The pieces are then glued and nailed or screwed together. A miter joint can be strengthened by inserting hardwood dowels or splines into the miter, as shown in *Figures 4-3c* and *4-3d*. You will need a router or drill press if you want to use such joints.

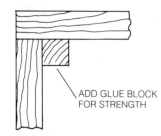
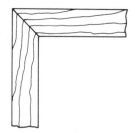
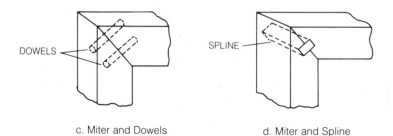

Figure 4-3. Common joint types are the butt and 45-degree miter. Miter joints can be strengthened by adding hardwood dowels or splines.

A few exotic joints can be used in building a speaker enclosure. They require special tools, such as dado blades and a router, as well as advanced carpentry skills. Consult a book on woodworking for details on how to make these special joints.

Some speaker enclosures are made so the front and rear panels fit into grooves made in the inside of the top, bottom, and side pieces, as shown in *Figure 4-4*. The width of the groove depends on the thickness of the front and rear panels. For a tight fit, most woodworkers make the groove slightly under-sized. Remember, the front and rear panels must be a little wider for this method as opposed to the method where the panels do not fit in grooves.

You can make the groove with a dado blade attached to a power saw, or by making successive cuts into the wood with a power saw and standard blade. A router can be used in some instances, but you must be sure to use a guide to ensure a straight cut. Ordinarily, the depth of the groove is 1/3 to 1/2 the thickness of the wood. You set the depth by adjusting the blade or bit in the cutting tool.

Sanding and Smoothing

All the surfaces of the wood should be sanded prior to assembling the pieces. A medium- to fine-grit sandpaper can be used. You should purchase an assortment pack with several grits. Start with the medium, 150 grit paper, and work your way to the fine, 200 grit paper. Don't bother trying to sand smooth the rough edges of particle board. You can clean up the cuts by sanding with a coarse sandpaper, but you will never get the edges smooth.

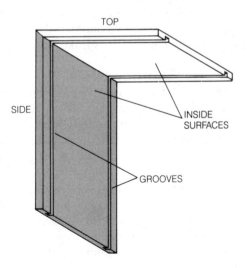

Figure 4-4. Grooves can be cut into the interior faces of the sides and top for holding the front and back pieces.

Aluminum oxide sandpaper lasts longer and does a better job than the cheaper flint or garnet sandpapers. Use a sanding block or motorized sanding tool to make the work go faster. A small file or rasp may be used to clean out the insides of dowel holes and grooves. Steel wool usually is not used when building speaker enclosures. The danger is that tiny threads of steel may remain after construction. They may cause an electrical short or be pulled into the space between the cone and voice coil by the speaker's magnet.

When you are finished sanding, remove the dust with a tack cloth (available at hardware stores), or use an ordinary clean cloth to brush away the dust. Be sure to get it all. Don't use a household dusting spray. After cleaning, inspect the pieces for physical damage and fill the voids with wood putty to make repairs. If you are using plywood, use wood putty to fill voids in the edges. After the putty dries, sand smooth.

When joining the pieces together to make butt joints, the ends of the wood will show. If the wood is stained, the end grain will absorb the stain much more readily than the surface, making the ends considerably darker. You can avoid this with several techniques. If you are staining the enclosure, apply molding trim or veneer to the ends. Use a trim or veneer that matches the wood you are using. If you are painting the enclosure, cover the ends with a thick coating of spackle or wood putty. Chapter 5 has more information on applying trim and veneer.

ASSEMBLING THE ENCLOSURE

Dry fit everything together to make sure the joints mate. If they do not, use a file or rasp to smooth the edges. You can test the fit of a joint by holding up the wood to the light. When the pieces are pressed together, no light should show through. If there are noticeable cracks, fix them or your speaker enclosure will not assemble as well as it should, nor will the speaker sound as good as it could.

Remove sanding dust and sawdust with the tack cloth once more before assembling the pieces. The usual steps of assembly are shown in *Figure 4-5*. Start by gluing and nailing (or screwing) the left side to the top piece. You can use white glue, but wood glue gives better results. Check the squareness of the joint with a square. Battens or glue blocks, shown as optional in *Figure 4-5*, not only strengthen the joint, but also help keep it square. More about battens is given later.

Use clamps to set the joints. Place scrap wood or cloth between the clamp and finished piece so the clamps don't mar the wood. Apply just enough pressure to bring the joints together, but not so much that all the glue oozes out. Keep the clamps on for at least one hour, preferably two or three, until the glue has dried completely. Do one joint at a time. If necessary, spread the assembly over a couple of days.

With most speaker enclosure designs, either the front, back or bottom piece is left off until the speakers and internal wiring have been installed.

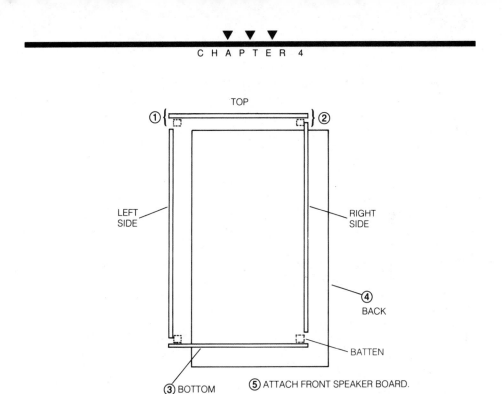

Figure 4-5. Typical construction procedure: 1) attach the top to the left side, 2) attach the top to the right side, 3) secure the bottom piece, 4) add the back, and 5) attach the front.

Using Nails and Screws

In most cases, the strongest joints are obtained using screws. But if nails are used, be sure to use the proper size for the thickness of wood you are using. Refer to *Table 4-2* for a listing of nail sizes and their use in speaker enclosures. If you do not plan to cover the wood with veneer or laminate, nail from the inside, or use a nail set to drive the nail heads below the surface. Use wood putty to fill the holes over the nail heads.

Table 4-2. Nail Size Chart

Type	Size	Length	Use
Common	3d	1¼″	Battens, Blocks
Finishing	2d	1″	Trim, Front
Finishing	3d	1¼″	Trim, Back
Finishing	4d	1½″	Back, Front
Finishing	6d	2″	Sides

For the strongest joint, use support blocks or battens, like those shown in *Figure 4-6*, inside the enclosure. Apply glue to the batten when nailing or screwing it on. The batten is necessary when using wood that has a thickness of 1/2 inch or less.

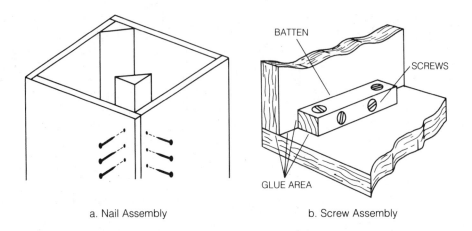

a. Nail Assembly   b. Screw Assembly

Figure 4-6. Battens or blocks on the inside of joints add strength. The extra wood pieces can be secured with nails or screws. Screws are recommended if the speaker enclosure is made from particle board.

Because particle board may flake apart when nailed, drill small pilot holes and assemble the enclosure with screws. Recommended screw sizes and lengths for various wood thicknesses are shown in *Table 4-3*. Glue the pieces together for a stronger bond. Clamping is not necessary because the screws keep the wood joint pressed together.

Table 4-3. Screws for Enclosure Construction

Type	Size	Length	Use
Flat head	#8	1¾"	Speaker Mounting Board
Round Head	#6	½"	Crossover
Round Head	#6	¾"	Tweeter/Midrange
Pan Head	#8	¾"	Woofer
Pan Head	#6	½"	Terminal strip. L-pad, etc.

The speaker mounting board and rear panel are most often attached to battens installed on the inside front edges of the enclosure, as shown in *Figure 4-7*. (If a grille board will be used in front of the speaker mounting board, be sure to leave space for it). The board is screwed in place. For best appearance, use a countersink or a special woodscrew bit to drill holes for the screws. The screw heads will be flush to the speaker mounting board and they will be easier to install. Any good carpentry book provides the details on how to use these special bits, and how to accurately gauge the proper bit to use with a particular screw size.

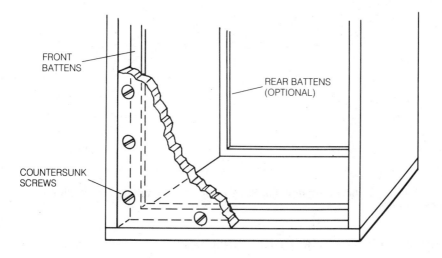

Figure 4-7. Battens hold the front and back pieces firmly inside the enclosure. They allow you to construct the enclosure without cutting grooves to hold the front and back pieces in place.

Fitting the Speakers

Depending on the design of the speaker enclosure, you will need to fit the speakers, wiring, and other electrical items before completing the enclosure. In some speaker designs, however, the back of the enclosure is installed last, so you complete the enclosure, including the front, before final assembly.

In any case, you need to fit the speakers to the mounting board. If you have not done so already, mark the wood for the speaker mounting holes. Use the drilling template included with the speaker, or use the speaker itself. Place the speaker or template flat against the wood, as shown in *Figure 4-8*. Use a pencil to mark the placement of the mounting holes.

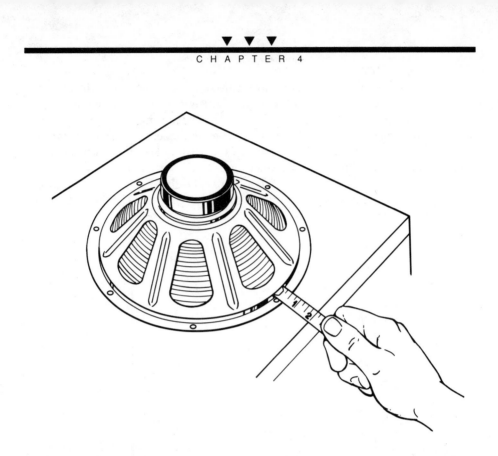

Figure 4-8. Use the mounting holes in the speaker as a template for marking the wood for drilling.

Drill the holes using the proper size bit. Machine bolts can be used to secure the speaker to the front panel, or you can use wood screws. Black-anodized hex-socket screws provide a high-tech look, but they are harder to find and more expensive than regular bolt and screw hardware. You can also paint the screw heads to match the speaker cone or frame. Whatever method you use, be sure that the speakers are mounted firmly. The enclosure will rattle if the speaker vibrates in its mounting. If the speaker comes with a mounting gasket, by all means use it.

Better sound dispersion is realized when the speakers are mounted so that the flange is on the front, not the back, of the speaker mounting board, as indicated in *Figure 4-9*. Always try for this configuration, especially if you are using a grille board in the enclosure. For best results, the grill board should be about the same thickness as the thickness of the speaker frame. This calls for a grille board 1/4 inch thick or less. The grille cloth must not touch any part of the cone or suspension.

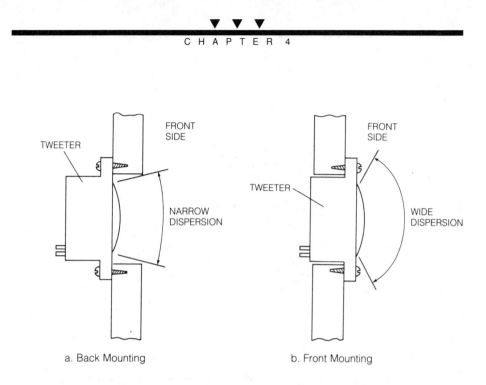

a. Back Mounting   b. Front Mounting

Figure 4-9. Speakers (especially tweeters) mounted from the rear of the speaker mounting board suffer from narrow dispersion. Always try to mount the speakers from the front.

Caulking the Joints

The slightest gap in the joints of your speaker enclosure can degrade the sound quality of system. You can seal the enclosure by using caulking, which is available in tube form at most home improvement stores. Caulking designed for tile grout is a good choice, or you may use the flexible silicone type.

Following the manufacturer's directions, apply the caulking to all inside joints of the enclosure. Allow the caulking to set completely. If you use battens to hold the speaker mounting board, do not caulk the front edge of the batten. Otherwise, the speaker mounting board will not rest firmly against the batten.

Adding Damping Material

Most speaker designs use damping material on the inside of the enclosure. You can use cotton or poly batting, available at most fabric stores, or custom-made speaker damping cloth such as 1 inch thick fiberglass material. Attach the damping material to the sides and bottom of the enclosure with a staple gun, as shown in *Figure 4-10*. Use 5/16 inch staples.

If you are using fiberglass, wear protective gloves and a dust mask. Tiny pieces of fiberglass can shake loose from the material, and may cause irritation to your skin and lungs.

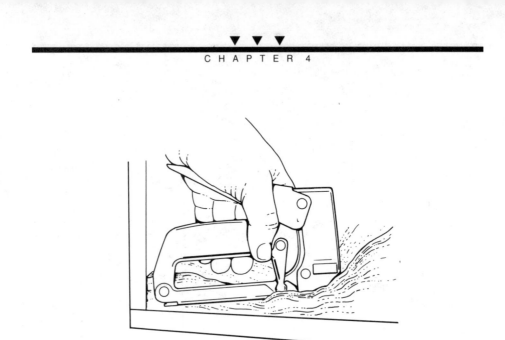

Figure 4-10. Use a staple gun to attach the damping material to the inside of the enclosure.

Remember that damping affects the sound reproduction of your speakers. In addition, it effectively alters the internal dimensions of the speaker enclosure. If you use damping you must take it into consideration when planning your speaker enclosures. Read Chapter 3 for more information on speaker damping and enclosure size.

INSPECTING YOUR WORK

Your speaker enclosures are almost finished. Before going on to the next step, inspect your work inside and out for the following irregularities:
- Gaps in the joints.
- Insufficient caulking of joints.
- Loose or unstable joints.
- Loose or poorly mounted speakers
- Loose damping material.
- Nails and screws not fully set into the wood.
- Gouges and hammer marks in the wood.

If you are making basic, unfinished enclosures, you can skip to Chapter 6 and learn how to wire the speakers for connection to your stereo system. Details for finishing the speaker enclosures — painting, varnishing, waxing, and veneering — are given in Chapter 5.

ADDING THE FINISHING TOUCHES

Unfinished, bare box enclosures may be fine for some rooms, but speakers for the living room and office need a more refined look. A few dollars and a few minutes time are all it takes to add the finishing touches to your homemade speaker enclosures.

This chapter shows you how a little bit of paint, varnish, veneer, or even contact paper can be used to greatly enhance the look of your speaker enclosures. We provide basic instructions for finishing your speaker enclosures, but you may wish to consult a book on carpentry for more complete details.

PAINTING

Painting is perhaps the easiest method of finishing your speaker enclosures. You can use any color that matches your decor. Prior to painting with the finish coat, brush or spray on a primer/sealer coat. This is particularly necessary when the speaker enclosure is made from particle board.

Primers take from 12 to 24 hours to dry; make sure the primer has completely dried before applying the finish coat. Finish coats will adhere better if the primer coat is sanded lightly before the finish coat is applied. Be sure to wipe the dust from the surface before painting with the finish coat.

LACQUERING

If you cannot find the color of paint you want in a gloss or semi-gloss finish, use a flat finish paint, then overcoat it with a clear gloss lacquer. The newest lacquers are designed to be applied over most paints, even enamels and acrylics. Read the instructions on the lacquer can to make sure.

VARNISHING AND STAINING

If the enclosure is made from veneered plywood or board planks, you can stain and/or varnish to protect and beautify the wood. Stains can be used separately from varnish; however, mixtures of stain and varnish are available so that both operations are done with one application.

The rub-on water-based varnish or stain is perhaps the easiest to use. However, the richest, deepest colors and grains are obtained by using oil-based stain, brushed on and then wiped off with a cloth. An oil-based varnish or clear urethane coating provides an excellent finish. Two or three coats, thoroughly dried and sanded in between, does the job. A few varnishes come in aerosol spray cans, but these are not recommended.

Some woods, like walnut and oak, are porous and need a light filler coat if you expect the finish to be glassy smooth. Most hardware and paint stores have a wide selection of fillers to choose from.

Waxing deepens the natural luster of varnished or stained wood. Any good paste wax designed for use on furniture can be used on your speaker enclosures. Apply several coats of the wax and polish each coat to brilliance.

APPLYING VENEER

Particle board and low-quality plywood can be covered with wood veneer. Just about every type of hardwood is available in veneer form. The thickness of the veneer varies, depending on the quality of the material and the type of wood. Much of it is 1/24 inch to 1/42 inch thick, and is rolled up like a paper blueprint.

Applying veneer is not difficult, but it does require patience and planning. The veneer is cut to size and shape using a sharp utility knife (remember to watch grain orientation). For best results, cut the veneer 1/16 to 1/8 inch larger than the part of the speaker enclosure you are covering. Lay the veneer good side up on a piece of cardboard or artboard. Use a straight edge and pencil to lightly draw the cutting line. With a metal straight edge or carpenter's square, score the veneer several times with the knife. Go slowly and don't slip, or you'll ruin the veneer.

Once the veneer is cut, it can be applied to the outside surfaces of the enclosure. If you are veneering the bottom of the speaker (which is not necessary), start there first; otherwise, start with the left or right side. Apply contact cement liberally to the surface of the enclosure and the under surface of the veneer. Use a good paint brush to apply it evenly.

Wait approximately 10 to 20 minutes for the cement to set (check the wait time in the product instructions), then apply the veneer to the enclosure. Be careful! The contact cement is extremely tacky and will permanently bond the instant the veneer touches it. Placing a piece of heavy waxed paper between the enclosure piece and the veneer can help you correctly position the veneer. When it is positioned as you want it, slip the waxed paper slightly so one end can bond. Then slowly slip the waxed paper out while pressing the veneer in place. Be sure that the veneer is properly aligned with the enclosure piece, or it will be forever crooked. Use a white rubber mallet to gently tap the veneer onto the enclosure, or use a piece of scrap wood with a smooth, flat surface as a buffer between a hammer and the veneer.

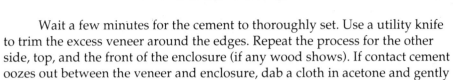
Wait a few minutes for the cement to thoroughly set. Use a utility knife to trim the excess veneer around the edges. Repeat the process for the other side, top, and the front of the enclosure (if any wood shows). If contact cement oozes out between the veneer and enclosure, dab a cloth in acetone and gently wipe away the excess.

FINISHING VENEER

You can apply finish to the veneer after the contact cement has fully dried. Sand the veneer with an extra-fine grit (220 to 280 grit) aluminum oxide sandpaper (be careful, the veneer is very thin). Remove the dust with a tack cloth. Apply stain and then varnish, or just apply varnish by itself. Certain woods, such as teak and mahogany, look exceptionally good after they have been rubbed with penetrating oil. The oils are available in a variety of colors.

As an alternative to veneer, you can apply wood-print contact paper (actually vinyl). The printing on the better quality papers looks remarkably like real wood, and they can be cut with an ordinary pair of scissors. If the paper doesn't have an adhesive backing, use contact cement to bond it to the speaker enclosure, as instructed for the veneer.

Since the paper is very thin and flimsy, it also must be applied slowly and carefully to prevent bubbles and wrinkles in the surface. A firm piece of plastic or metal (such as a paint shield) is very useful to work out all the bubbles and get a smooth surface.

ACOUSTIC GRILLES

Grille Frames

The acoustic grille cloth attaches to a frame or grille board and covers up the cones of the speakers. In some speaker designs, the grille frame or board is removable; in others it is permanent. It is far easier to make the grille removable, so we will use a removable grille for our design. A grille frame can be constructed out of hardwood trim or molding. Window- or door-frame molding is a good choice because it gives the grille cloth a gently curved edge. You install the molding so that the thinner edge faces outward.

Construct the frame by cutting 45-degree mitered corners. Glue the joints and secure with corrugated fasteners (used for picture frames), as shown in *Figure 5-1*. Grille frames for large speakers may require a cross brace. Position the brace so that it is between the speakers (between the woofer and midrange is a good location).

Grille Cloth

The best grille cloth is acoustically transparent but visually opaque. You can use most any fabric for the grille cloth, but you must be certain that it doesn't muffle the sound. Most fabrics will block at least a certain portion of the audio spectrum, which colors the sound. To block the view of the speakers, you may

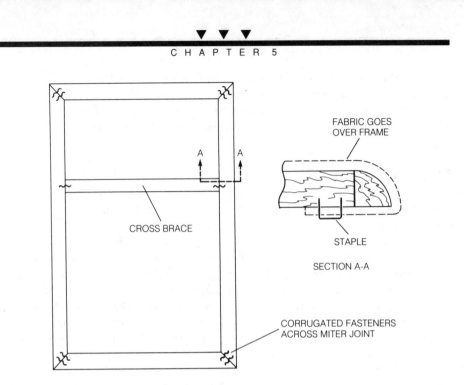

FABRIC GOES
OVER FRAME

STAPLE

SECTION A-A

CROSS BRACE

CORRUGATED FASTENERS
ACROSS MITER JOINT

Figure 5-1. The joints of the grille frame are glued and secured using corrugated fasteners.

need to apply several layers of the cloth, which increases the coloration of the sound.

The grille cloth in *Figure 5-2* is specifically designed for use with speaker systems. It not only blocks the view of the speakers, but also passes most of the sound spectrum. The grill cloth fits completely over the frame, covering all surfaces with fabric.

Attach the cloth to the frame using the following procedure: Make sure that the pattern or weave of the cloth is positioned correctly, then staple one corner of the cloth to the bottom of the frame. Work your way to the opposite corner while stapling the edges of the grille cloth to the bottom of the frame, as shown in *Figure 5-2*. One staple every one to two inches is sufficient. After stapling one edge, pull the cloth over the frame and staple the opposite edge. When you are finished, there should be no puckers in the grille cloth. At the corners, cut the cloth to fan it out and make a smooth surface. Trim the excess cloth and staple it in place. Make sure that no part of the grille cloth touches any moving part of the speaker. Finish the back of the frame by applying tape or cloth to hide the staples (see *Figure 5-2*).

Attach the frame to the speaker enclosure with fabric or plastic hook-and-loop strips. The hook part goes on the frame, the loop part goes on the speaker enclosure. Staple the hook part to the frame, using fabric glue to help hold it in place. Use contact cement or epoxy to fasten the loop part to the speaker enclosure. For best results, position a small 1 by 1 inch strip at each corner of the frame.

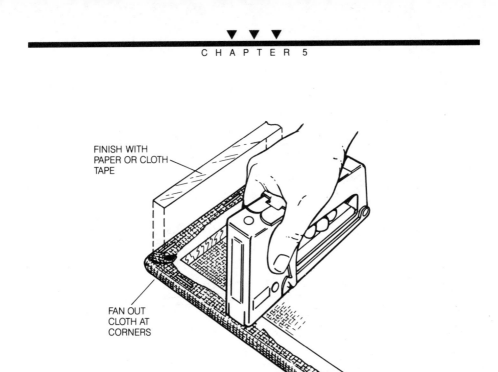

FINISH WITH
PAPER OR CLOTH
TAPE

FAN OUT
CLOTH AT
CORNERS

Figure 5-2. Staple the grille cloth to the back of the frame and make neat, square corners by carefully cutting and fanning the fabric. Finish with tape to hide staples.

Grille Board

A grille board is a piece of wood with cutouts for the speakers. The cutouts are larger than the ones in the speaker mounting board, so the holes in the grille board may run into one another, as some do in *Figure 5-3*. A 1/4 inch thick piece of plywood or particle board is ideal for use as a grille board.

After the board has been cut to size, and the holes made (following the procedures outlined in Chapter 4), you can attach the grille cloth. Staple the cloth to the back of the board, using the same methods as for the grille frame. Be sure to make smooth tucks at the corners. Again, make sure that the cloth does not touch any moving parts of the speaker.

Hook-and-loop strips can be used to attach the grille board to the enclosure; however, they are usually too weak for the heavy grille board and may allow it to vibrate. You must make sure it is attached securely so it doesn't vibrate. Screws provide a more secure attachment. Decorative screws, some with decorative washers or small escutcheons, can be obtained from a hardware store. Drill clearance holes in the board and pilot holes in the enclosure. Four to six screws should be used to attach the board solidly so it will not vibrate. The speaker board is sometimes recessed into the enclosure.

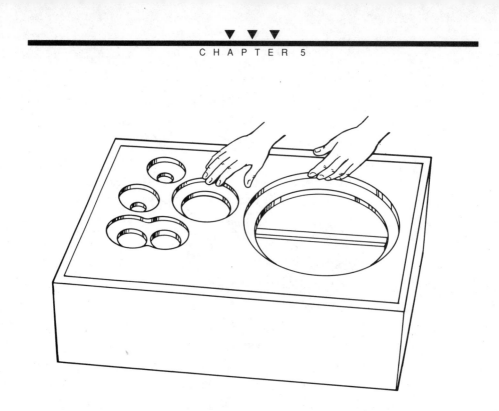

Figure 5-3. The grill board has slightly larger cut-outs for the speakers. The board adds thickness to the front end, so take it into consideration when attaching the speaker mounting board. Place the speaker mounting board further into the enclosure so that the grille board is flush with the front of the enclosure.

WIRING
YOUR
SPEAKER

Connecting your speakers to your hi-fi entails more than soldering a wire here or there. Your speakers must be connected properly if you expect them to deliver the rich, full sound they were designed to produce. Additional components are sometimes required to properly interface the speakers in the enclosure to the amplifier.

This chapter details the internal wiring of speakers. You'll learn how to connect a one-speaker system and a multiple-speaker system to a hi-fi, how crossover networks provide better sound and protect speakers against damage, how to install and use L-pads, and why you should use fuse protection.

BASIC ELECTRICAL WIRING

Single-speaker systems are the easiest to connect to an amplifier. Speaker or lamp cord is connected between the speaker terminals and the amplifier output terminals. Stereo systems require two speaker enclosures. A simplified schematic diagram of a stereo system wiring scheme is shown in *Figure 6-1*.

Wiring for more advanced speaker systems is based on the one-speaker layout. But no matter how complex the system is, there are several important considerations when wiring any speaker enclosure.

Wire Size

The louder the sound, the more power is sent through the wires. If the amplifier and speakers are designed for high output, you need large wires to prevent signal loss and heating in the wires. Wire size is commonly stated as a gauge. The commonly used gauge is the American Wire Gauge (AWG). The larger the gauge number, the smaller the wire diameter. *Figure 6-2* shows recommended maximum wire gauge numbers (minimum wire diameters) for interconnecting wire lengths for a typical hi-fi with a maximum of 100 watts per channel output. Note that the greater the distance between speaker and amplifier, the larger the wire size (smaller gauge number). Large wire (small gauge number) has lower resistance, so it has less voltage drop than a smaller wire. Even with the recommended wire sizes of *Figure 6-2*, there can be 10% to 20% loss at the maximum length. Use the next larger wire size to minimize line losses.

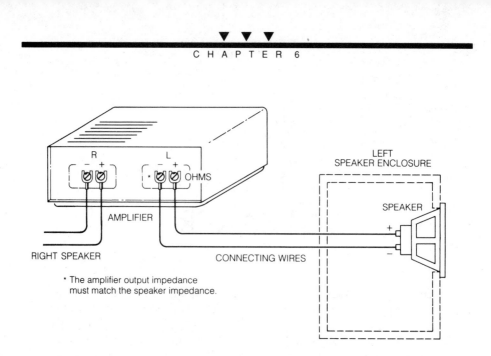

Figure 6-1. Basic speaker hookup entails connecting the output terminals of the amplifier to the terminals of the speaker. Be sure to observe polarity between the amplifier and speaker. A stereo amplifier hookup is shown. The speaker imped- ance must match the amplifier output impedance.

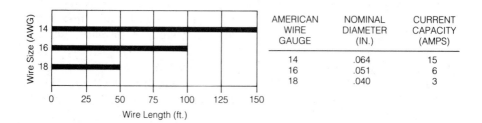

AMERICAN WIRE GAUGE	NOMINAL DIAMETER (IN.)	CURRENT CAPACITY (AMPS)
14	.064	15
16	.051	6
18	.040	3

Figure 6-2. The longer the distance between speaker and amplifier, the larger the wire diameter required.

Speaker Polarity

All speakers are polarized. You need to make sure that the positive and negative terminals on the amplifier are connected to corresponding terminals on the speaker. The positive and negative terminals on most speakers and amplifiers are identified by one of the following methods:

- A "+" symbol indicates positive; a "−" symbol indicates negative.
- A red dot or terminal indicates positive; a black dot or terminal indi- cates negative.

- The center terminal on the common RCA phono jack is usually positive and the outer terminal is usually negative.

If your speaker terminals are not marked for polarity, you can readily test them with a size D flashlight battery. *Momentarily* touch the speaker wires to the battery terminals while watching the movement of the cone. The cone moves only as the battery is connected or disconnected. Do not leave the battery connected to the speaker terminals.

- If the cone moves outward when the battery is connected, then the positive side of the battery is connected to the positive terminal on the speaker.
- If the cone moves inward when the battery is connected, then the negative side of the battery is connected to the positive terminal on the speaker.

Mark or label the polarity of the speaker terminals for future reference. Be careful when using this technique on small dome tweeters; you may damage them.

Speaker Enclosure Terminals

On simple speaker wiring jobs, you can solder the wires to the speaker terminals and lead the wires out the back of the enclosure. Drill a small hole in the enclosure back panel and thread the wires through it. If the speaker is designed for exterior use, you may want to drill the hole larger and insert a rubber grommet. The grommet helps prevent chafing of the wires. You can caulk around the wires or grommet to prevent air and moisture from leaking through the hole.

Elaborate speaker enclosures require external terminals to provide easy connection between the speaker system and hi-fi. A number of terminal styles are available; *Figure 6-3* shows some popular types. You need to make cutouts for the terminals in the back of the speaker enclosure. Use a saber saw, or if the cutout is small enough, a manual coping saw. Most terminals are flanged so that any rough edges of the cutout are hidden.

Inside the enclosure, solder wires to the speaker and enclosure terminals. Use 12- or 14-gauge flexible wire. Most terminals are marked for polarity.

Wiring Two or More Speakers

All speakers have impedance, which is a measurement of the resistance to alternating electrical current. The impedance of most high-fidelity speakers is eight ohms, though some speakers are rated at either four or 16 ohms.

When two or more speakers are interconnected in a single speaker enclosure, the impedance presented to the amplifier at the enclosure terminals is different than for a single speaker. No matter how the wiring is arranged, the impedance is affected. The impedance at the speaker enclosure terminals (or for connections to a single speaker) should match the output impedance of the amplifier.

a. Quick Connect Flush Mount

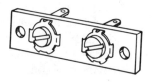
b. Screw Terminals

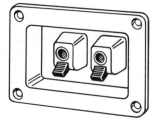
c. Quick Connect Recessed Mount

Figure 6-3. Popular types of speaker enclosure terminals.

- Wiring the speakers in series, as shown in *Figure 6-4a*, doubles the terminal impedance (assuming both speakers have the same impedance). If both speakers are rated at eight ohms, for instance, the system impedance is 16 ohms. For the same amplifier output voltage, one half as much power will be delivered to the speakers. To obtain the same power from the combination requires a larger voltage output from the amplifier.
- Wiring the speakers in parallel, as shown in *Figure 6-4b*, halves the terminal impedance (assuming both speakers have the same impedance). Total system impedance with two eight ohm speakers is only four ohms. For the same amplifier output voltage, twice the power is delivered to the speakers. The same power from the combination requires a larger current from the amplifier than for a single speaker.

CROSSOVER NETWORKS

Although you can wire two or more similar speakers together in one enclosure, you can't connect a tweeter, midrange, and woofer together, and expect the high, middle, and low frequencies to magically find the correct speaker. In fact, feeding high-power, low-frequency signals to an electrodynamic tweeter will quickly burn it out.

A crossover network divides the sound frequency spectrum into distinct ranges, and ensures that only the proper frequencies are routed to the appropriate speaker. You can make your own crossover networks, or buy them ready-made. Commercially-made crossover networks are available for both two- and three-way speaker systems.

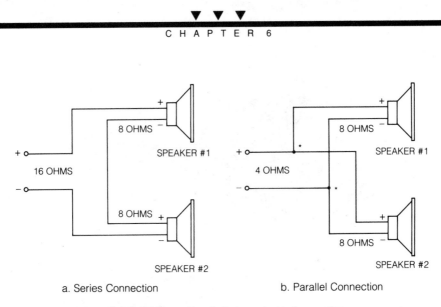

a. Series Connection   b. Parallel Connection

*Dots at line intersections indicate an electrical connection.
No dots means no electrical connection.

Figure 6-4. (a.) Two speakers with the same impedance wired in series causes the impedance to double. (b.) Two speakers with the same impedance wired in parallel causes the impedance to decrease to one-half.

The better commercial crossover networks, like the models shown in *Figure 6-5*, are easy to use, inexpensive, and can be easily tailored to match the speakers you are using. The networks are rated to 100 watts, which is enough for most hi-fi systems. All crossover networks are passive electrical devices; they don't require any operating power.

Using a Crossover Network

The specification sheet that accompanies most better-quality speakers indicates the effective frequency or dynamic range of the speaker.

Table 6-1 lists recommended crossover frequencies for different-sized woofers in both two- and three-way speaker systems. Most midrange speakers have a frequency range from about 1500 Hz to about 6000 Hz; tweeters are active beyond about 6000 Hz. You can obtain good results by using crossover frequencies that are within these ranges.

Note that the crossover frequency of the network and the frequency range of the speaker do not need to match exactly. A network with a crossover frequency of 1600 Hz can be effectively used with a woofer that has an upper frequency limit of 2000 Hz.

The two-way crossover is tapped to provide two crossover frequencies for the woofer and midrange/tweeter. Depending on how you connect the network, you may select 2000 Hz, 2500 Hz, or 4000 Hz for the woofer and midrange/tweeter crossover points. The input terminals (marked IN) connect to the amplifier.

The three-way crossover is also tapped to let you select the best crossover frequencies for the speakers you are using. You may choose either 800 Hz

a. Two-way Crossover Network

b. Three-way Crossover Network

Figure 6-5. Two popular crossover networks, for two-and three-way speaker systems. *(Courtesy of Radio Shack)*

Table 6-1. Crossover Frequencies for Woofers

Two-Way System	
Woofer Diameter (in.)	**Suggested Crossover Frequency (Hz)**
8	2500
10 to 12	1600
Three-Way System	
Woofer Diameter (in.)	**Suggested Crossover Frequency (Hz)**
8	1000/5000
10 to 12	700/4500
15	600/4500

or 1600 Hz for the woofer, 800-5000 Hz or 1600-7000 Hz for the midrange, and 5000 or 7000 Hz for the tweeter.

Note that you should avoid connecting the speakers so that there are "holes" in the frequency response. For example, don't connect a woofer to the 800 Hz terminal and the midrange to the 1600-7000 Hz terminal. That leaves the frequency range between 800 and 1600 Hz open.

Figures 6-6 and *6-7* show how to connect the two- and three-way cross-over networks to make a working speaker system. You may choose different connecting points depending on the speakers you use.

Networks that lack multiple taps for each speaker are wired in a similar manner, but you have no way of choosing optimum crossover points. The terminals are marked for the proper speaker: "T" is for tweeter, "MR" is for midrange, and "W" is for woofer. The negative terminal on each speaker connects to the common or "–" (minus) terminal of the crossover network.

Using a Two-Way Crossover with Three Speakers

Just because you use a two-way crossover doesn't mean you can't build a three-way speaker system. *Figure 6-8* shows how to wire a woofer, midrange, and tweeter to a two-way crossover network. Use the capacitor value shown for a crossover frequency of approximately 9000 Hz. You may delete the capacitor if you are using a piezoelectric tweeter.

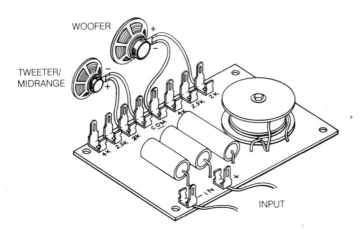

Figure 6-6. Connecting a woofer and tweeter to the 2 kHz crossover points on a two-way crossover network. *(Courtesy of Radio Shack)*

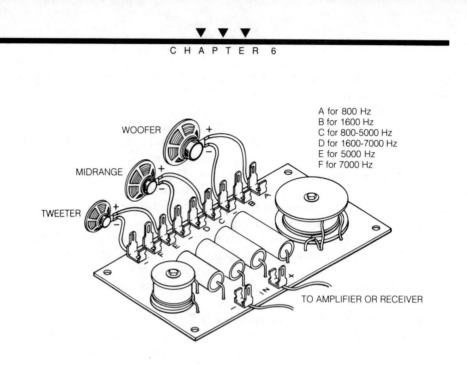

A for 800 Hz
B for 1600 Hz
C for 800-5000 Hz
D for 1600-7000 Hz
E for 5000 Hz
F for 7000 Hz

Figure 6-7. Connecting a woofer, midrange, and tweeter to a three-way crossover network. The woofer uses the 800 Hz crossover point; the midrange uses the 800-5000 Hz crossover point; and the tweeter uses the 5000 Hz crossover point. *(Courtesy of Radio Shack)*

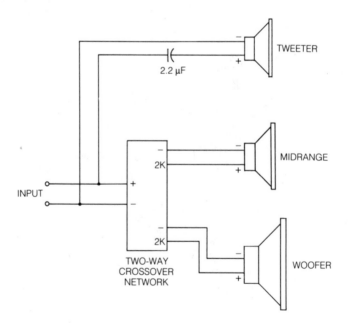

Figure 6-8. Connecting three speakers with a two-way crossover network.

Installing the Crossover Network

Ready-made crossover networks come assembled on printed circuit boards. The boards can be mounted directly to the inside of the speaker enclosure. An ideal mounting location is on the inside of the speaker driver board. If there is no room on the driver board, secure the crossover network to the inside of the back panel. Holes are pre-drilled on the network circuit board for easy mounting. Use washers or additional nuts to space the circuit board away from the wood of the enclosure.

Observe correct terminal polarity when connecting speakers to a crossover network.

INSTALLING L-PADS

It's inevitable that one speaker in the enclosure will sound "brighter" than the others. Tweeters and midranges are usually more efficient than woofers, so the highs may be unnaturally enhanced. You can adjust the volume level of each speaker in the enclosure by installing fixed resistors, but it is difficult to determine the proper values. An L-pad, like the one in *Figure 6-9*, permits you to adjust the output of the speakers in the enclosure so that the tone is more pleasing to your ear. Just by turning a knob, you can bring out the highs or accentuate the lows. L-pads alter the sound output by changing the response of the speaker system.

An L-pad is connected between the speaker and the amplifier if a crossover network is not used. If you have installed a crossover network, the L-pad is connected between the network and the speaker. The L-pad maintains a constant impedance load for the amplifier or network while varying the signal level to the speaker. All L-pads have three terminals, usually numbered 1, 2, and 3. These numbers are used in the wiring diagrams in *Figure 6-10*.

Figure 6-9. An L-pad, capable of handling up to 75 watts of audio power.
(Courtesy of Radio Shack)

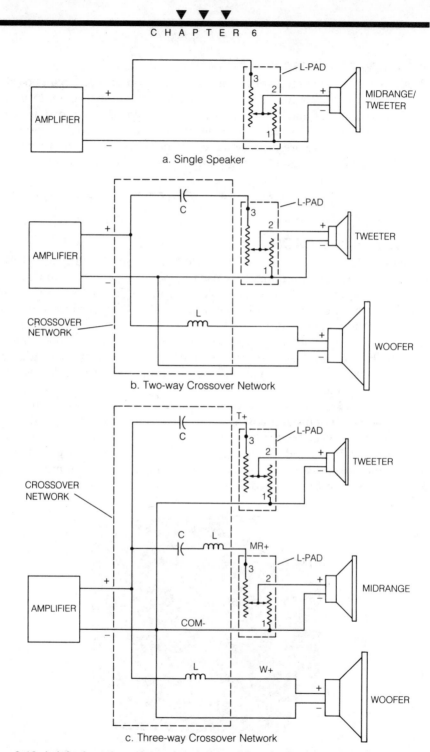

a. Single Speaker

b. Two-way Crossover Network

c. Three-way Crossover Network

Figure 6-10. (a.) An L-pad used as a level control for single speaker. (b.) One L-pad in-line with a tweeter and two-way crossover network. (c.) Two L-pads in-line with a midrange, tweeter, and three-way crossover network.

Figure 6-10a shows the basic wiring for an L-pad used in a full-range single-speaker system. (In this application, the L-pad serves as a level control rather than a tone control). *Figures 6-10b* and *6-10c* show L-pads used in a two-way and three-way speaker enclosure, respectively. In the latter two figures, the L-pads are used as a tone balance control.

As with speakers, L-pads have a maximum wattage rating, usually 50 or 75 watts. Excessive power levels can burn out the wire coils inside the control. You can safely use 75-watt L-pads with most any system. The 25-watt models are designed for connection to tweeters only, or for use in low-power audio applications.

L-pads are usually mounted on the driver board, next to the speakers they control. If you are installing L-pads in your enclosures, plan for them when you cut the holes for the speakers. To mount the control, drill a hole large enough for the shaft to fit through. The front panel of the control is flanged, and has mounting holes for 6-32 or 8-32 hardware. For L-pads that require a cutout, use a saber saw or coping saw.

Don't secure the L-pad to the driver mounting board if the speaker grille is not removable, or if the knob would protrude beyond the front of the enclosure. In these instances, mount the control on the rear panel of the speaker enclosure.

FUSE AND DYNAMIC OVERLOAD PROTECTION

Fuses and dynamic overload protectors should be installed in your speaker enclosure systems to prevent damage to the speakers. A dynamic overload protector, wired as shown in *Figure 6-11*, helps prevent premature burn-out of your tweeters. Use one protector for each tweeter in the system. Overload protectors are normally soldered in place as near to the positive terminal of the tweeter as possible.

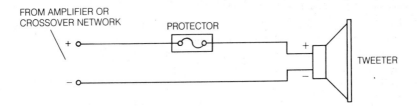

Figure 6-11. Connect the overload protector in-line with the positive terminal of the tweeter.

Install a speaker fuse holder on the outside rear panel of the enclosure, or connect it in-line between the amplifier and speaker (or crossover network), as shown in *Figure 6-12*. Install a two to ten amp fuse in the holder.

Use the following formula to determine the proper rating for the fuse:

$$A = \sqrt{P/Z}$$

where: A is the current rating of the fuse in amps
P is the power handling capacity of the speaker in watts
Z is the impedance of the speaker in ohms.

Insert values for P and Z, perform the division, and find the square root of the result. For example, a 75 watt, eight ohm speaker system requires a three amp fuse. Calculator keystrokes are: clear, 75 ÷ 8 = 9.38, square root, gives 3.06.

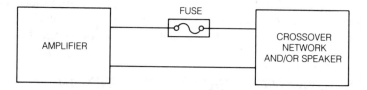

Figure 6-12. Connect the fuse in-line with the positive terminal of the speaker or crossover network.

FINAL ASSEMBLY AND CHECKOUT

With the speaker wiring complete, you can finish the assembly of the enclo-sure. With most enclosure designs, this means attaching the bottom or rear panel. Be sure the wires don't touch the speakers or the speaker cones. If they do, they can cause the speakers to emit a loud and annoying rasp.

In the next chapter, you'll learn how to connect your speaker enclosures to your hi-fi and how to test them for proper operation. You'll also learn how to place the speakers in the room for optimum sound quality.

PUTTING YOUR SPEAKERS TO WORK

Success with your speakers depends on how well they are integrated with the rest of your hi-fi system. A haphazard arrangement will yield poor results, denying you the full capabilities of your speakers.

Careful wiring and placement of your speakers enhances their sound quality, and makes listening more pleasurable. This chapter tells you everything you need to know about wiring your speakers to your hi-fi system. It also tell you how to test them for proper operation and how to place them in the room for optimum listening enjoyment.

CONNECTING SPEAKERS TO YOUR AMPLIFIER

Use two-wire speaker cable to connect your speaker enclosures to your hi-fi system. Follow the guidelines given in Chapter 6 for wire size. As mentioned there, too small a wire size could restrict you from being able to enjoy the full power and capability of your speakers. When in doubt, use a larger wire diameter (smaller gauge) to avoid power loss.

Connection Procedure

Follow these simple steps when connecting your speakers to your audio amplifier:

1. Always turn off the amplifier before attaching any cables or wires to it. If you don't, you may short out an output from the amplifier and damage it.
2. Measure the length of cable needed between the amplifier and speaker. Cut cables for both left and right speakers the same length, even if you place one enclosure next to the amplifier. You reduce the chance of channel imbalance by maintaining even cable lengths. Try to keep the cables as short as possible, but allow yourself some slack so you can freely position the speakers anywhere in the room.
3. Separate the conductors on each end of the two-wire cable to about two to three inches. Strip approximately 1/4-inch of the insulation off each conductor end, as shown in *Figure 7-1*. Take care not to nick or cut off any strands of the conductors. Twist the ends of the wires to prevent the strands from fraying.

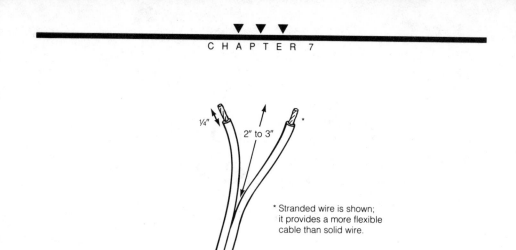

Figure 7-1. Strip 1/4" of insulation from the ends of the wires. For best results, tin the wire ends with a small amount of solder.

4. Note the polarity key of the wire. Most speaker cables are keyed for polarity — either a raised edge or a different color for one conductor. If the cable is not keyed, use an ohmmeter as shown in *Figure 7-2* to identify the conductors.

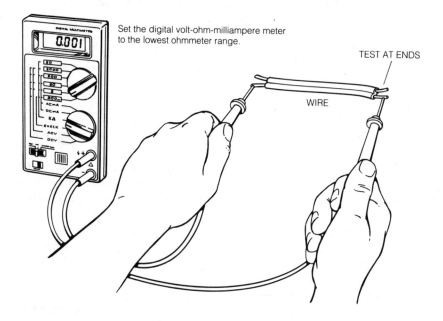

Figure 7-2. Use an ohmmeter to identify the conductors of the speaker wire. Set the ohmmeter to measure low resistance.

5. Attach both conductors at one end of the wire to the speaker. Connect the keyed conductor to the positive terminal. *Figure 7-3a* illustrates connections to quick connect terminals. If your speakers have screw terminals, wrap the end of the wire around the screw shank in a clockwise direction, as shown in *Figure 7- 3b*. Be sure that none of the wires fray out and touch the other terminal. Tinning the end of the wires with a small amount of solder keeps the strands together and provides a neat connection. Alternatively, you may use solderless crimp lugs. Attach the lugs to each end of the wire and insert the lug under the screw head of the terminal.

6. Attach both conductors at the other end of the wire to the amplifier. Be sure to connect the keyed conductor to the positive terminal. Some amplifiers have more than one set of speaker terminals, usually identified "A" and "B." Connect the first pair of speakers to the "A" set of terminals. Connect the second pair, if used, to the "B" set of terminals. Stereo systems have a left (L) and right (R) output and may have "A" and "B" speakers for each output.

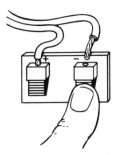

Insulation or wire characteristics identify the + conductor.

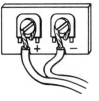

a. Quick Connect Terminals

b. Screw Terminals

Figure 7-3. (a.) Press the terminal buttons to insert the wire. After you release the button, pull lightly on the wire to be sure it is secure. (b.) Loop the wire around the screw terminal at least one turn to ensure a solid connection.

Cable Routing

If you're like most hi-fi enthusiasts, you probably have a tangle of wires behind your stereo system. For better sound, spend a few moments to "dress" the wires to your speakers so that they are neater. This not only makes troubleshooting problems easier, it reduces the chance of ac-induced hum entering signal cables and being amplified by the system. Keep these points in mind:

- If the wire is too long, coil it in a large loop and secure the coil with tape or a plastic tie wrap.
- Whenever possible, separate speaker wires and ac cords by several inches. When this isn't possible, cross speaker wires and ac cords only at a 90 degree angle; never run the two in parallel. Otherwise, noise from the ac lines will be induced in the speaker wires.

- Avoid coiling ac cords, because this creates a strong magnetic field, like an electromagnet or motor, that can induce hum in nearby signal wires.
- Never route the cables where they can be stepped on or tripped over. Avoid running the wires under carpeting, except along the wall.

Checkout

Turn down the volume on the amplifier, then apply power to it. If your amplifier has a speaker select switch, set it to "A" or "B" in accordance with the terminals you used to attach to the new speakers. Select a function (CD, cassette, phono, etc.) and start the music. *Slowly* turn the volume up to a comfortable listening level.

If your speakers are connected to the amplifier properly, and all the internal wiring is correct, you should hear sound. Should you not hear anything, check that the music is really playing (watch the level meter or display on the amplifier, or use a pair of headphones). If that doesn't cure the problem, refer to *Table 7-1*, which lists possible causes and remedies for speaker malfunctions. It assumes that the amplifier is working properly.

TESTING

Once you have determined that the speakers are working, you should now take a few moments to make sure that they are working properly. The first tests involve your ears only; the last test requires the use of a sound level meter.

Speaker Function Test

Purpose of Test: To determine if all of the speakers in the enclosure are operating properly.

1. Turn on a sound source on your hi-fi (record, cassette, compact disc, or FM radio station are good choices). Make sure all power switches are on.
2. Turn the amplifier balance control so that only the left speaker enclosure is on. If both speaker enclosures remain on, there is a short in the speaker terminals on the amplifier. Be sure that no strands of wire cross and interfere with adjacent terminals.
3. Adjust the volume so that the listening level is comfortable when you are next to the speaker.
4. Make sure all level controls (L-pads) are full on (full clockwise position). Place your ear six to eight inches from each speaker in the enclosure. You should hear sound from each speaker. The sound from the tweeter should be relatively high; the midrange and woofer speakers should emit medium and bass tones, respectively. If this is not apparent, change to another musical selection and perform the listening test again. Should the sound not change, there may be a problem in the internal wiring in the speaker or in the crossover network.
5. Repeat the process for the right speaker.

Table 7-1. Basic Troubleshooting Guide for Speaker Installation

Problem	Cause	Remedy*
No sound	Speaker not properly connected to amplifier.	Check wiring for shorts and opens.
	Amplifier function not set correctly.	Select proper music source.
	Speaker not selected.	Set amplifier speaker select switch.
	Speaker defective.	Replace speaker.
Weak sound	Volume control down.	Turn up volume control.
	Speaker wires shorted or have broken strands.	Check wiring.
	Speaker wire too small.	Use larger wire (smaller gauge).
	Speaker defective.	Replace speaker.
Buzzing sound	Speaker wiring short.	Check wiring.
	Damping material or grille cloth touching speaker cone.	Check for material touching speaker.
	Speaker defective.	Replace speaker.
No balance control	Terminals shorted at amplifier.	Check wiring.
	Stereo/Mono switch set to Mono.	Set switch to Stereo.
Excessive highs or lows	Incorrect crossover wiring.	Check wiring.
	Incorrect values for crossover components.	Check values.
	Speaker out of circuit.	Check that all speakers are working.
Fuse blown	Speaker overdriven.	Reduce volume and replace fuse.
	Short in wiring, crossover or speaker.	Check wiring and components.
	Fuse value too low.	Replace with correct ampere rating.
	Speaker defective.	Replace speaker.

*These remedies assume that the amplifier is working properly.

Left/Right Speaker Polarity Test

Purpose of Test: To determine if the left and right speaker enclosures are wired to the amplifier with the same polarity.

1. Place the speakers side by side, at a distance of about eight feet away from you.
2. Put on a music selection with a strong bass line.
3. Adjust the volume to a comfortable listening level.
4. Listen to the music for 10 to 15 seconds, and note the level of bass. You should not be able to easily identify the sound coming from either speaker; rather, the speakers should sound as one unit.

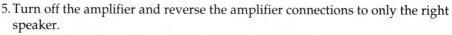
5. Turn off the amplifier and reverse the amplifier connections to only the right speaker.

6. Turn on the amplifier, and again listen to the music for a few seconds, and note the bass.

7. If the bass now sounds weaker, the speakers were wired correctly in the first place. Return to the original wiring arrangement.

8. If the bass now sounds stronger, leave the speakers connected as they are.

L-Pad Test

Purpose of Test: To determine if the L-pads are working properly.

1. Tune in a FM station.

2. Turn the volume control on the amplifier to a low level. Set the tone controls (bass/treble or frequency equalizer) to the middle of their range.

3. Turn off the "FM mute" on the tuner or receiver. Tune to a spot on the FM dial where there is no station so you hear only hiss. Turn up the volume if the hiss is difficult to hear.

4. Turn the balance control so that only the left speaker is on.

5. Adjust the L-pad(s) while listening for a change in sound. If you don't hear a change, the L-pads are not wired correctly.

6. Turn the balance control so that only the right speaker is on. Repeat step 5.

Sound Level Test

Purpose of Test: To determine the frequency response of the speakers and adjust for inconsistencies. For this test you need a sound level meter and a test record (or tape). Test records are available at most large record stores and through mail order (check the ads in hi-fi magazines).

1. Position the sound level meter on a tripod or stand approximately three feet from the front of the speaker and at least three feet from the floor. If you don't have a tripod or stand, you can have an assistant hold the meter, but take care that the microphone of the meter is not blocked, and that the meter isn't moved during the test.

2. Set the weighting switch to C. Set the range switch to 80 dB. Set the response switch to Slow.

3. Place the test record on the turntable and adjust the volume control to a comfortable listing level. Set the tone or equalization controls to the middle of their range. Select the 1000 Hz tone track on the record, and note the reading on the sound level meter.

4. Adjust the volume control on the amplifier so that the meter reads 0 dB. If you can't get the sound level meter to read 0 dB, set the meter range switch to another number.

5. Change record tracks to select other frequencies, and note the meter reading. On a piece of graph paper, plot the reading at each frequency on the test record.

6. When you finish plotting, analyze the graph. It won't be perfectly flat, but you should not see dips or peaks that are more than about five dB either above or below the 0 dB reference level.

7. You can attempt to flatten the response by adjusting the L-pad(s) on the speaker, altering the equalization at the amplifier, or changing the speaker position.

SPEAKER PLACEMENT

You can ruin the sound from even the best hi-fi by carelessly placing the speakers in the corners of the room. It's best to consider the listening room as an extension of the speaker enclosure. In fact, the rich, full sound you hear from a good set of speakers comes not only from the speaker enclosure, but also from the sound reverberating throughout the room.

When placing your speakers, feel free to experiment until you achieve the sound *you* like best. Don't be afraid to try something daring and different. The following pointers should start you in the right direction.

Room Considerations

Some room sizes are better than others. Your speakers may sound much better in your den than in your living room. Apart from room acoustics caused by windows, drapes, carpeting, and other surfaces, the room dimensions play an important role.

Speakers almost always sound better when they are positioned along the long wall of the room. If the sound is too boomy, the speaker placement in that particular room may be causing what's known as standing waves. Try a new location.

Avoid placing the speakers in the corner of the room. This greatly diminishes smooth frequency response because the sound bounces off three surfaces, and produces heavy bass.

It's best to position the speakers two to three feet from all walls. The closer a speaker is placed to a wall, the more bass it produces.

If the stereo image is blurred or lacks depth, there may be excessive middle and high frequency reflections off the walls. Damp these by placing a wall hanging or other soft, absorbent material on the wall between the two speakers.

Speaker Height

As you learned in previous chapters, the sound from the tweeters in your speaker enclosures does not disperse as well as from the midranges and woofers. Therefore, placing the speakers so that the tweeters are at about ear level provides the smoothest response. This usually means placing the speakers off the floor, either on a set of speaker stands (you can buy them or make your own) or hanging the speakers on wall brackets.

If you are hanging speakers on the wall, use mounting brackets designed for the job, and be sure to use the proper anchoring hardware. Avoid placing the speaker too close to the floor or ceiling, where it will unnaturally boost the bass range.

Speaker Separation

Place speakers six to eight feet away from each other. Placing them closer together diminishes the stereo effect; too far apart creates a sonic "hole" in the stereo effect.

Aim at the Sweet Spot

Position the speakers so that they are directed to the central listening point in the room. That point shouldn't be any closer than about eight feet from the speakers. If closer, the direct sound from the speakers will overpower the room acoustics. The spot where the sound from the two speakers meet is called the "sweet spot" (*Figure 7-4*), and is where the stereo effect is the greatest. Position the right and left speakers so that they are the same distance away from the sweet spot. This increases the stereo effect. Experiment a bit, because if you toe-in the speakers a bit more so that their axes cross in front of your favorite listening spot you will often expand the area over which good stereo is obtained.

FOUR SAMPLE SPEAKER DESIGNS

If you haven't done so already, you're probably itching to build your first speaker system. Using the information contained in this and previous chapters, you should have no trouble in designing and constructing your own high-quality home-made speaker systems. For your convenience, Chapter 8 contains plans for four ready-to-go speaker systems. All dimensions and components are listed, but feel free to modify the enclosures to suit your tastes and requirements.

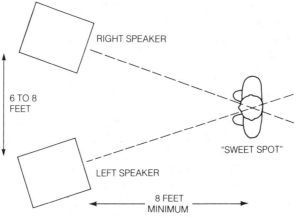

Figure 7-4. The "sweet spot" is the area where the stereo image is the strongest.

PROJECT PLANS

You don't need to do a lot of calculations and computations to make your first speaker enclosures. This chapter presents four plans you can use for your own speaker enclosures, spanning from simple "bookshelf" speakers designed for a small den or child's room, to large living room sets meant to be used with a high-power amplifier. Each plan includes a full parts list of lumber, hardware, speakers, and miscellaneous items. *Table 8-1* shows a comparison of enclosure specifications for the four plans. The relationship of volume, maximum wattage, and Q of the box are included. The text indicates hole sizes by drill bit number; for your convenience, *Table 8-2* is a cross reference between drill bit numbers and their corresponding fractional inch sizes.

SYSTEM 1 — COMPACT BOOKSHELF SPEAKER

System 1 has a small volume of 0.3 cubic feet, a Q of 1.1, and a maximum power handling capacity of 20 watts. The compact bookshelf speaker enclosure is designed for use in small dens, in a child's room, in mobile homes, and in other places where space is at a premium or where you don't want (or need) a large speaker system. Despite its small size, this compact bookshelf speaker has a warm, pleasing tone that complements most any musical taste. For ease in construction, the enclosure has few components and just one full-range 6-inch speaker. *Table 8-3* gives a parts list.

The plans presented here are for a speaker enclosure made of particle board and covered with wood grain contact paper. You can finish the speaker using any technique you desire, including painting, laminating with veneer, or staining (when constructed with cabinet-grade plywood).

Table 8-1. Speaker Enclosure Specifications

System	Volume cu. ft.	Q	Maximum wattage
System 1	0.30	1.1	20
System 2	0.52	1.0	50
System 3	3.32	1.5	50
System 4	2.15	0.8	50

Table 8-2. Drill Bit Cross Reference

Bit Number	Fractional Inch Size
48	$5/64$
33	$7/64$
28	$5/32$
19	$11/64$
10	$3/16$

Table 8-3. Parts List for System 1

Qty.	Part	Description
1	Top	¾-inch particle board; 8½ by 6½ inches
1	Bottom	¾-inch particle board; 8½ by 6½ inches
2	Sides	¾-inch particle board; 12 by 6½ inches
2	Battens	Scrap ¾-inch particle board or 1 by 1-inch pine; 5 inches long
1	Speaker/grille	⅜-inch plywood; 13½ by 8½ inches
1	Back	⅜-inch plywood; 12 by 7 inches
1		6-inch full range speaker
1		10½ by 15½ inch grille cloth
4		No. 6, 1-inch wood screws
4		6-32, ¾-inch machine bolts, nuts
Misc.		3d finishing nails, wood glue, caulking, 25 feet 18 or 20 gauge speaker wire, fiberglass damping material, contact paper.

Construction Steps

1. Cut the front driver board (which also serves as the speaker grille board) and back from 3/8-inch plywood as indicated in *Figure 8-1*. Cut the top, bottom, and sides from 3/4-inch particle board.
2. Sand the surfaces of the wood with medium-grit sand paper. Small chips in the edges of the wood can be filled later before applying the contact paper.

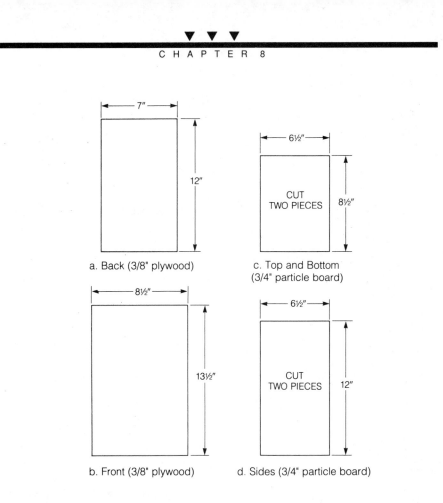

Figure 8-1. Cutting layout for System 1.

3. Cut four 5-inch battens using scrap pieces of particle board. Set two aside for later. In two of the battens, drill a pair of pilot holes, using a #48 bit, all the way through the batten, spaced about 1½ inches from either end. These will be used when mounting the driver board. Glue and nail two of the battens centered between the side edges and flush to the front of the bottom and of the top pieces of the enclosure. The holes you drilled should face the front and back of the enclosure.

4. Assemble the top, bottom, and sides as shown in *Figure 8-2*, with the side pieces sandwiched between the top and bottom. Spread glue around the joints and nail together using 3d nails.

5. Use a rasp or motorized sander to apply a slight chamfer to the front four edges of the front piece. The chamfer provides a sloping surface for the grille cloth and enhances the look of the enclosure.

6. Mark the speaker hole in the front piece, following the layout guide in *Figure 8-3a*. Be sure to cut the hole smaller than the flange of the speaker, or there won't be enough wood left to mount the speaker to the board. Use a compass, string, or stick, as detailed in Chapter 4, to mark the circle for the

hole (see *Figure 8-3b*). To start the cutout, drill a hole *inside* the line with a 3/8-inch bit. Insert the blade of a hand-held saber saw into the hole as shown in *Figure 8-16b* and begin sawing. Don't worry about a perfectly round cut-out; no one will see it once the enclosure is complete.

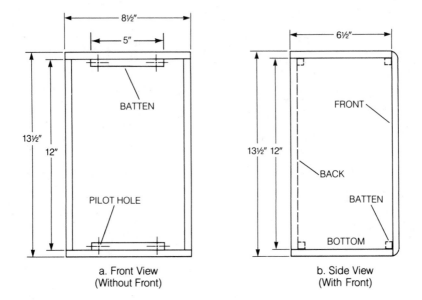

a. Front View
(Without Front)

b. Side View
(With Front)

Figure 8-2. Assembly details for System 1.

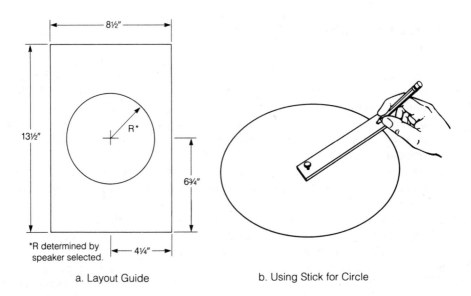

*R determined by speaker selected.

a. Layout Guide

b. Using Stick for Circle

Figure 8-3. Driver board layout for System 1.

7. Mark the locations for the four mounting screws of the speaker and drill holes using a #19 bit. Countersink the holes on the front of the driver board so that the mounting bolts are flush. Solder a 25 foot (or so) length of 18 or 20 gauge speaker wire to the terminals of the speaker. Tie a loose knot in the wire about one foot from the speaker.

8. Mount the speaker as detailed in *Figure 8- 4a* using 6-32, 3/4-inch bolts and nuts.

9. Stretch a piece of 10½ by 15½-inch grille cloth over the front, wrap around edges, and staple it to the back of the front panel as indicated in *Figure 8-4b*. Remove excess cloth and tape down the edges using paper tape (see Chapter 4).

10. Place the driver board and attached speaker over the front of the box. Invert the box and insert No. 6, 1-inch wood screws into the four pilot holes you previously drilled in the mounted battens. Tighten the screws into the driver board.

11. Apply caulking to the inside of the enclosure, being sure to seal the cracks between all the joints. Don't get any caulking on the speaker (see *Figure 8-13a*).

12. Secure the remaining two battens to the inside top and bottom of the box 3/8 inch from the back. Center them between the two side edges. Use glue and nails for a permanent bond.

13. Using a #19 bit, drill a hole in the back panel near the bottom and thread the speaker wire through it. The knot you tied in the wire should be positioned to prevent strain on the speaker's solder connection if the wire is yanked.

14. Stuff about 0.5 cubic feet of acoustic fiberglass damping inside the enclosure, being careful not to crush the speaker by jamming the insulation against it.

15. Place the back panel in the enclosure, flush to the battens. If you've cut the wood properly, the back should just fit into the box. If the back is too small, cut another piece. Conversely, if the back is too large, trim it with the saw or wood rasp. Using a #48 bit, drill two holes each along the top and bottom through the back piece and battens. Use No. 6, 3/4-inch wood screws to fasten the back panel in place. You may, if you wish, apply a light bead of wood glue around the back panel joints.

Finishing the Enclosure

An easy way to finish the enclosure is to apply wood-grain contact paper to the sides, top, and bottom. Start by cutting the paper to seven inches wide, then line up the end of the paper to one of the bottom corners with about 1/8 inch overlap to the front. While peeling the backing off the paper, wrap it around the sides and top of the enclosure (see *Figure 8-5*). Work slowly and remove buckles and bubbles with your fingers.

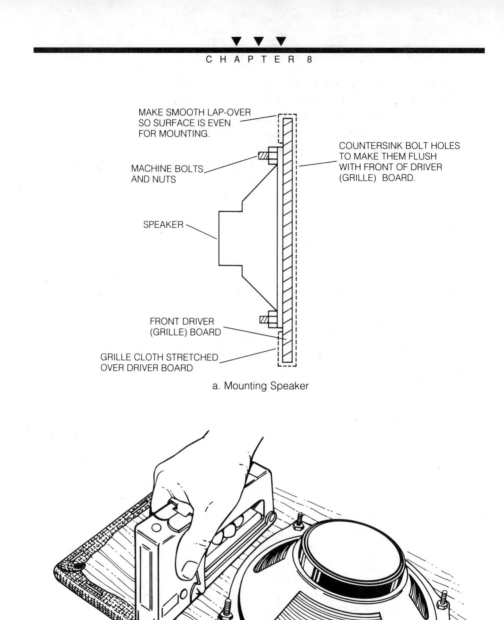

MAKE SMOOTH LAP-OVER
SO SURFACE IS EVEN
FOR MOUNTING.

COUNTERSINK BOLT HOLES
TO MAKE THEM FLUSH
WITH FRONT OF DRIVER
(GRILLE) BOARD.

MACHINE BOLTS
AND NUTS

SPEAKER

FRONT DRIVER
(GRILLE) BOARD

GRILLE CLOTH STRETCHED
OVER DRIVER BOARD

a. Mounting Speaker

b. Stapling Grille Cloth

Figure 8-4. Mount the speaker flange behind the driver board using machine bolts and nuts, then staple on grille cloth.

Figure 8-5. Smooth the contact paper over the surface.

Trim the excess paper in the front of the enclosure and tuck it in the small crack between the box and front speaker grille board. As shown in *Figure 8-6,* cut miters at the back corners and carefully fold the 3/8 inch overlap of the paper over the back. This should produce a nice finished look over the back joints.

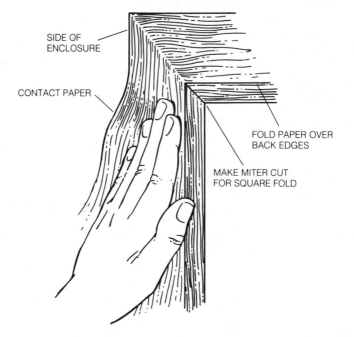

SIDE OF
ENCLOSURE

CONTACT PAPER

FOLD PAPER OVER
BACK EDGES

MAKE MITER CUT
FOR SQUARE FOLD

Figure 8-6. Fold and trim the contact paper to make a smooth corner.

SYSTEM 2 — SMALL LIVING ROOM/DEN 2-WAY SPEAKER

System 2 has a volume of 0.52 cubic feet, a Q of 1.1, and 50 watts maximum power. This two-way enclosure is small enough to fit in tight spaces, such as a living room bookshelf, yet because of its two speakers, delivers full, rich sound. An L-pad is provided so you can tailor the tone of the speaker to suit your listening tastes. Though this project uses two speakers and one L-pad, it is designed without a separate crossover network. Instead, it uses a single capacitor on the tweeter (the capacitor comes with the speaker specified in the parts list). Alternatively, you can wire a 2-way crossover network following the procedure given in Chapter 6.

Construction Steps

1. Refer to *Table 8-4* for the parts list and to *Figure 8-7* for the cutting layout. Cut the lumber to size (top, bottom, two sides, back, and driver board). Although the parts list calls for 3/4-inch plywood, you may use 3/4-inch particle board if you plan to paint or cover the enclosure with contact paper, wood laminate, or veneer. Note that the top and side pieces are mitered, as

Table 8-4. Parts List for System 2

Qty.	Part	Description
1	Top	¾-inch plywood; 12 by 7 inches
1	Bottom	¾-inch plywood; 10½ by 7 inches
2	Sides	¾-inch plywood; 17 by 7 inches
4	Battens	Scrap ¾-inch plywood or 1 by 1-inch pine; 10½ inches long
1	Driver board	¾-inch plywood; 10½ by 15½ inches
1	Back	¾-inch plywood; 10½ by 15½ inches
2	Grille frame	1 by 1-inch pine; 12 inches long
2	Grille frame	1 by 1-inch pine; 17 inches long
2	Grille frame (support)	1 by 1-inch pine; 10½ inch long (approx., measure to fit, as braces)
1		8-inch woofer speaker
1		Dome tweeter, with capacitor
1		25-watt peak L-pad
1		14 by 19 inch grille cloth
1		Speaker terminals
8		No. 6, ½-inch wood screws (for speakers)
4		No. 6, ½-inch wood screws (for L-pad)
4		No. 6, 2-inch wood screws (for attaching driver board and grille)
2		6-32, 1-inch machine bolts with nuts (for terminal)
Misc.		3d and 4d finishing nails, corrugated fasteners, wood glue, caulking, 10 feet 16 or 18 gauge speaker wire, fiberglass damping material, wood stain.

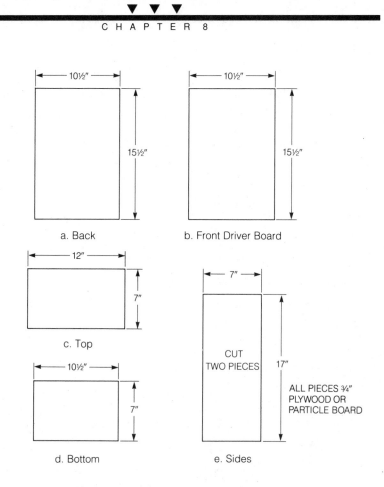

Figure 8-7. Cutting layout for System 2.

shown in the assembly drawings of *Figure 8-8*. Adjust the blade of your table or circular saw to make a 45 degree cut. Try the setting with a piece of scrap wood. When the miters are butted together, the angle must be exactly 90 degrees. Use a carpenter's square to assure precision.

2. Assemble the top, sides, and bottom as indicated in *Figure 8-8* using 4d nails. Strengthen the joints by adding short batten pieces using scrap left over from the 3/4-inch plywood or use 1 by 1-inch pine. The mounting edge of the battens should be placed 3/4 inch in from the front and back surfaces of the top and bottom pieces. Place them so they do not interfere with the placement of the other pieces. You'll find it easier to secure the battens using wood screws and glue, because the small interior of the enclosure makes it hard to swing a hammer.

3. Sand the surfaces of the wood with medium-grit sand paper.

4. This enclosure has a grille cloth stretched over a grille frame. This is then mounted to the front of the enclosure. Construct the grille frame using 1 by 1-inch pine. Cut the pieces as indicated in the parts table. Miter the ends as shown in *Figure 8-9*. Cut two more pieces of 1 by 1-inch pine to a length of

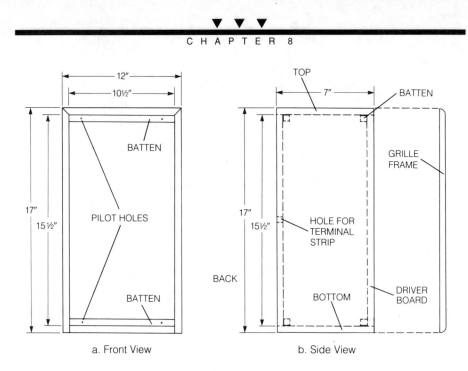

a. Front View   b. Side View

Figure 8-8. Assembly details for System 2.

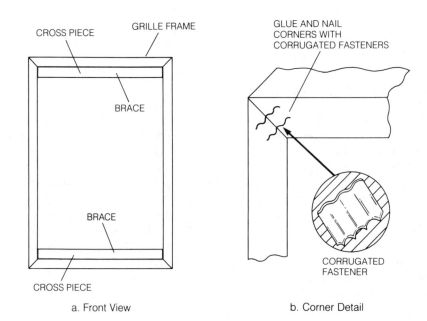

a. Front View   b. Corner Detail

Figure 8-9. Details of the grille frame.

10½ inches. Secure them to the underside of the top and bottom frame pieces to brace the frame assembly. Apply glue to the corners and fasten all the pieces using 3d nails or corrugated fasteners. Stretch a 14 by 19-inch piece of grille cloth over the frame and secure it using staples (see Chapter 4 if you are not sure how to do this). Set the grille frame aside.

5. Mark the holes for the speaker and L-pad in the driver board, following the layout guide in *Figure 8-10* and the layout drawings for the speakers. After cutting out the holes, temporarily insert the speakers and L-pad to be sure that they fit. Enlarge the hole, if necessary, to accommodate the components.

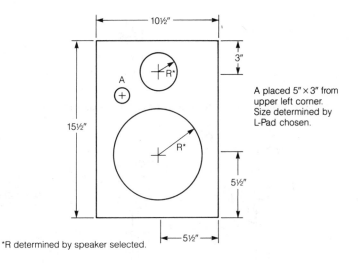

*R determined by speaker selected.

A placed 5" × 3" from upper left corner. Size determined by L-Pad chosen.

Figure 8-10. Driver board layout for System 2.

6. Mark the locations for the speaker and L-pad screws and drill holes using a #33 bit. (See finishing instructions if you wish to stain the driver board before assembly.) Mount the speakers as detailed in *Figure 8-11* using No. 6, 1/2-inch or No. 8, 1/2-inch wood screws. Use No. 6, 1/2-inch wood screws to secure the L-pad to the driver board.

7. Wire the speakers, L-pad, and terminal as shown in the wiring diagram in *Figure 8-12*. Be sure to allow enough wire length to mount the speaker terminal strip in the center of the back panel.

8. In the two front battens, drill a pair of pilot holes, spaced about 1½ inches from either end, using a #48 bit. The holes you drilled should face the front and back of the box.

9. With glue along the board edges, place the driver board into the front of the box. Redrill the pilot holes in the battens using them as guides to drill pilot holes into the driver board. Mount the grille frame to the front of the box. Insert No. 6, 2-inch wood screws into the four holes you previously drilled and tighten them into the driver board and the grille frame.

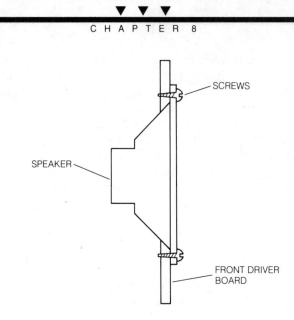

Figure 8-11. Mount the speaker flange in front of the driver board.

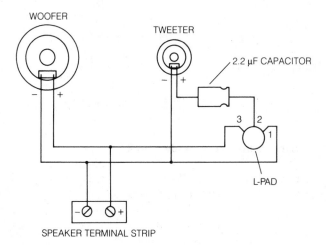

Figure 8-12. Wiring diagram for System 2 speakers and L-pad.

10. Apply caulking to the inside of the enclosure, being sure to seal the cracks between all the joints (see *Figure 8-13a*). Don't get any caulking on the speaker.
11. Cut a hole in the center of the back panel, using the screw terminal strip as a template. Secure the terminal strip using two 6-32, 1-inch machine bolts and nuts.
12. Staple 1-inch acoustic fiberglass damping to the sides and bottom of the box (see *Figure 8-13b*). Staple a single sheet to the inside of the back panel.

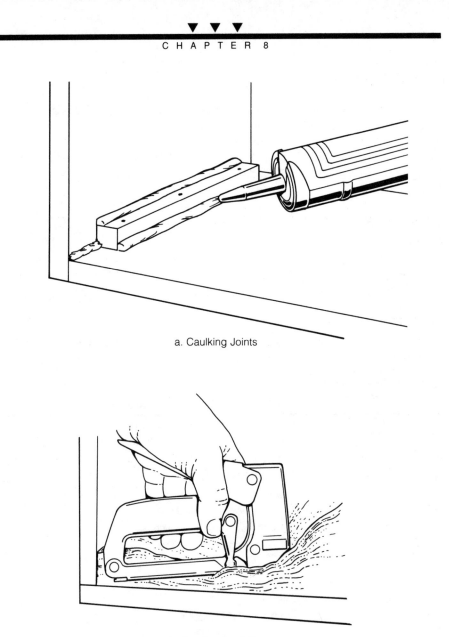

a. Caulking Joints

b. Stapling Insulation

Figure 8-13. Caulk the joints and add insulation.

13. Place the back panel inside the enclosure. If you've cut the pieces properly, the back should just barely fit into the box. When pushed against the back battens, it should be flush with the back of the sides of the enclosure. Secure the back panel with six No. 6, 1½ -inch wood screws into the battens (three on the top and three on the bottom). Apply a light bead of wood glue around the outside joints of the back to provide a good seal.

Finishing the Speaker

Sand the enclosure one more time using fine-grit sand paper. Fill any gaps, chips, cracks, and voids with wood filler or putty. After the filler is dry (allow at least four hours), apply stain to the wood using a brush-on or wipe-on technique. You may wish to stain the enclosure before final assembly. Specifically, stain the enclosure before you attach the driver board and grille frame. Staining the driver board with a dark color prevents the wood from showing through the grille cloth.

SYSTEM 3 — LARGE LIVING ROOM 3-WAY SPEAKER

System 3 is a three-way speaker with a volume of 3.32 cubic feet, a Q of 1.5, and a maximum power of 50 watts. The plans show you how to construct an acoustic suspension 3-way speaker enclosure. The enclosure uses a powerful 12-inch woofer suitable for high output amplifiers.

Construction Steps

1. See *Table 8-5* for the parts list and *Figure 8-14* for the cutting layout. Cut the lumber to size.

Table 8-5. Parts List for System 3

Qty.	Part	Description
1	Top	¾-inch plywood; 19½ by 13¼ inches
1	Bottom	¾-inch plywood; 18 by 13¼ inches
2	Sides	¾-inch plywood; 29¾ by 13¼ inches
4	Battens	1 by 1-inch pine; 18 inches long
4	Battens	1 by 1-inch pine; 27½ inches long (approx.; measure to fit)
1	Driver board	¾-inch plywood; 18 by 29 inches
1	Back	¾-inch plywood; 18 by 29 inches
1	Grille board	⅜-inch plywood; 19½ by 30½ inches
1		12-inch woofer
1		4-inch midrange, soft domed;
1		Dual radial horn tweeter
1		3-way crossover network
1		18 by 29-inch grill cloth
1		Speaker terminal
16		No. 6, ½-inch wood screws (for speakers)
8		No. 6, 1¼-inch wood screws (for attaching driver board)
8		No. 6, 1½-inch wood screws (for back panel)
2		6-32, 1-inch machine bolts with nuts (for terminal)
4		No. 6, ½-inch decorative screws for grille board
Misc.		3d and 4d finishing nails, corrugated fasteners, wood glue, caulking, 10 feet 16 gauge speaker wire, fiberglass damping material, wood stain

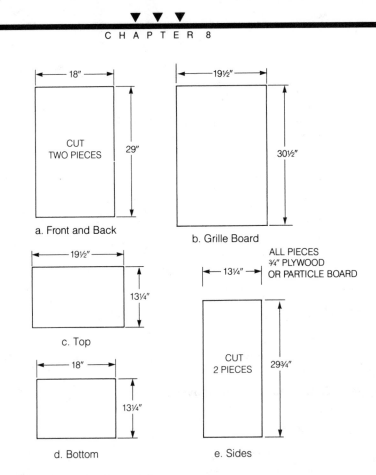

Figure 8-14. Cutting layout for System 3.

2. Assemble the pieces as indicated in *Figure 8-15* using 4d nails. Strengthen the side to top and side to bottom joints by adding short 3-inch battens as shown. You can use either scrap lumber left over from the 3/4-inch plywood or use 1 by 1-inch pine. Position the battens so that they do not interfere with the the other pieces.

3. Mark the speaker holes in the driver board, following the layout guide in *Figure 8-16a*. After cutting the holes (see *Figure 16b*), temporarily insert the speakers to be sure they fit. Enlarge the hole, if necessary, to accommodate the components.

4. Using the cutouts in the driver board as a guide, mark the holes in the 3/8-inch grille board. Cut the wood with a saber saw, taking an extra 1/4 inch to 3/8 inch (some of the holes may run into one another, as shown in *Figure 8-17*, but this is acceptable for the grille board). Stretch an 18 by 29-inch piece of grille cloth over the frame and secure it using staples. Set the grille board aside.

5. Mark the locations for the speaker mounting holes in the driver board and drill pilot holes using a #33 bit. Mount the speakers using No. 6, 1/2-inch or No. 8, 1/2-inch wood screws.

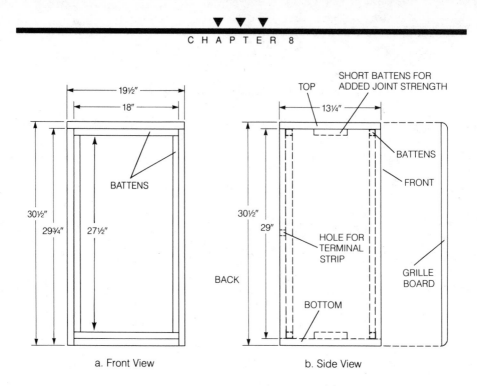

a. Front View

b. Side View

Figure 8-15. Assembly details for System 3.

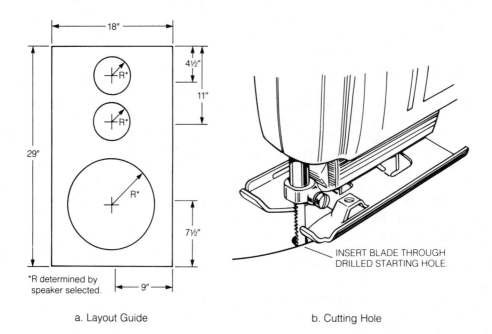

a. Layout Guide

b. Cutting Hole

Figure 8-16. Driver board layout for System 3.

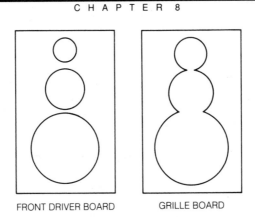

FRONT DRIVER BOARD   GRILLE BOARD

Figure 8-17. Cut slightly larger holes for the speaker openings in the grille board.

6. Wire the speakers, crossover network, and terminal as shown in the wiring diagram in *Figure 8-18*. Be sure to allow enough wire length to mount the speaker terminal strip in the center of the back panel.

7. Cut four 18-inch and four 27½-inch long battens using scrap pieces of plywood. In two 18-inch battens and two 27½-inch battens, drill a pair of pilot holes, spaced about 1½ inches from the ends (use a #48 bit). Glue and nail the four battens 1⅛ inches from the front top, bottom, and sides of the enclosure. The holes you drilled should face the front and back of the box.

8. Place the driver board over the front of the box. Insert No. 6, 1¼-inch wood screws in the previously drilled holes and tighten them into the driver board.

9. Apply caulking to the inside of the enclosure, being sure to seal the cracks between all the joints. Don't get any caulking on the speaker (see *Figure 8-13a*).

10. Mount the other set of four battens 3/4 inch in from the rear opening on the top and bottom of the box.

11. Cut a hole in the center of the back piece, using the screw terminal strip as a template. Secure the terminal strip using two 6-32, 1-inch machine bolts and nuts.

12. Staple 1-inch acoustic fiberglass damping to the sides and bottom of the box. Staple a single sheet to the inside of the back panel (see *Figure 8-13b*).

13. Place the back panel inside the enclosure. The back should fit snugly into the enclosure and seat firmly against the four battens. Secure the back panel with eight No. 6, 1½-inch wood screws tightened into the battens. Apply a light bead of wood glue around the outside joints in the back to provide a good seal.

Finishing the Speaker

Finish the speaker following the procedure given for System 2. When you are finished, slip the grille board into place over the driver board. Attach the grille board with four decorative No. 6, 1/2-inch wood screws. You may have to drill pilot holes in the driver board.

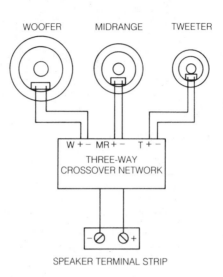

WOOFER   MIDRANGE   TWEETER

W + − MR + −   T + −
THREE-WAY
CROSSOVER NETWORK

SPEAKER TERMINAL STRIP

Figure 8-18. Wiring diagram for System 3 speakers.

SYSTEM 4 — PORTED REFLEX 3-WAY SPEAKER

This advanced project uses an 8-inch woofer and ported reflex design for accentuated bass. Two L-pads let you selectively alter the response of the speakers.

1. Refer to *Table 8-6* for the parts list and to *Figure 8-19* for the cutting layout. Cut the lumber to size. Note the joints that are mitered, as shown in the assembly diagram in *Figure 8-20*. Adjust the blade of your table or circular saw to make a 45 degree cut. Try the setting with a piece of scrap wood. When the miters are butted together, the angle must be exactly 90 degrees.

2. Assemble the pieces as indicated in *Figure 8-20* using 4d nails. Strengthen the joints by adding short, 3- or 4-inch batten pieces to the top and bottom side joints using scrap left over from the 3/4-inch plywood. Alternatively, you can use 1 by 1-inch pine for the battens.

3. Construct the grille frame as indicated in *Figure 8-21* using 1 by 1-inch pine. Cut to the size given in the parts list. Miter the ends as shown in *Figure 8-21* and add a cross brace in the middle for stability. Apply glue to the corners and fasten all the pieces using 3d nails or corrugated fasteners (see *Figure 8-9b*). Stretch an 18 by 30-inch piece of grille cloth over the frame and secure it using staples.

4. Mark the speaker holes in the driver board, following the layout guide in *Figure 8-22*. After cutting out the holes, temporarily insert the speakers and L-pads to be sure they fit. Enlarge the holes, if necessary, to accommodate the components.

5. Mark the mounting screw holes for each speaker and L-pad and drill them using a #33 bit. Mount the speakers using No. 6, 1/2-inch or No. 8, 1/2-inch wood screws. Use No. 6, 1/2-inch wood screws to secure the L-pads to the driver board.

Table 8-6. Parts List for System 4

Qty.	Part	Description
2	Top and Bottom	¾-inch plywood; 15½ by 11½ inches
2	Sides	¾-inch plywood; 28 by 11½ inches
4	Battens	1 by 1-inch pine; 14 inches long
4	Battens	1 by 1-inch pine; 26½ inches long (approx.; measure to fit)
1	Driver board	¾-inch plywood; 14 by 26½ inches
1	Back	¾-inch plywood; 14 by 26½ inches
2	Grille frame	1 by 1-inch pine; 15½ inches long
2	Grille frame	1 by 1-inch pine; 28 inches long
2	Grille frame (support)	1 by 1-inch pine; 14 inches long (approx., frame measure to fit)
1		8-inch woofer
1		Wide dispersion soft-dome midrange
1		1¾-inch dome tweeter
1		3-way crossover network
1		18 by 30 inch grille cloth
1		Speaker terminal
2		75 watt peak L-pads
1		Fuse holder
1		Fuse; 1 amp
1		2-inch black ABS plastic pipe; 1¼ inches long
12		No. 6, ½-inch wood screws (for speakers)
8		No. 6, 1¼-inch wood screws (for attaching driver board)
4		No. 6, ½-inch wood screws (for L-pad)
8		No. 6, 1½-inch wood screws (for back panel)
4		6-32, 1-inch machine bolts with nuts (for terminal)
Misc.		4d finishing nails, corrugated fasteners, wood glue, caulking, 10 feet 14 or 16 gauge speaker wire, fiberglass damping material, hook and loop fasteners, wood stain

6. Cut a piece of 2-inch black ABS plastic pipe to 1¼ inches. Use a file to smooth the rough edges. Use the pipe as a template and trace around it with a pencil at the position for the port shown in *Figure 8-22*. Cut out the hole, being careful to make it as circular as possible. Fit the pipe into the driver board flush with the front and apply caulk around the inside edges to secure it and seal it in place.

7. Wire the speakers, L-pads, fuse, crossover network, and terminal strip as shown in the wiring diagram in *Figure 8-23*. Use wires long enough to reach each component. Be sure there is plenty of wire to reach the terminal strip mounted in the center of the back panel.

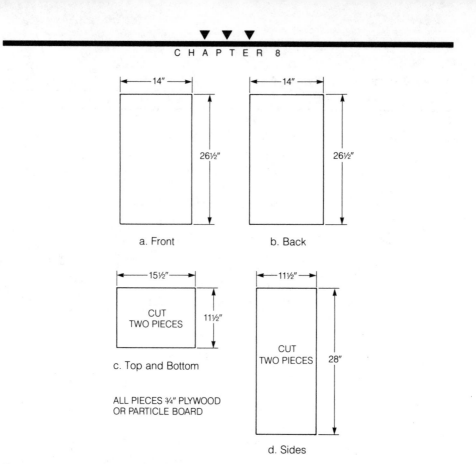

a. Front

b. Back

CUT
TWO PIECES

c. Top and Bottom

ALL PIECES ¾" PLYWOOD
OR PARTICLE BOARD

CUT
TWO PIECES

d. Sides

Figure 8-19. Cutting layout for System 4.

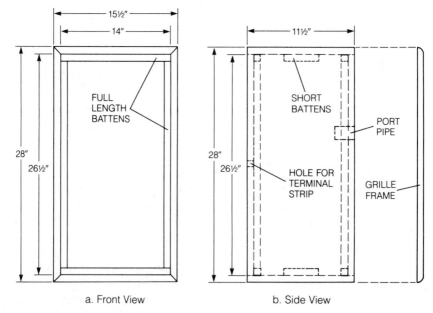

FULL
LENGTH
BATTENS

SHORT
BATTENS

PORT
PIPE

HOLE FOR
TERMINAL
STRIP

GRILLE
FRAME

a. Front View

b. Side View

Figure 8-20. Assembly details for System 4.

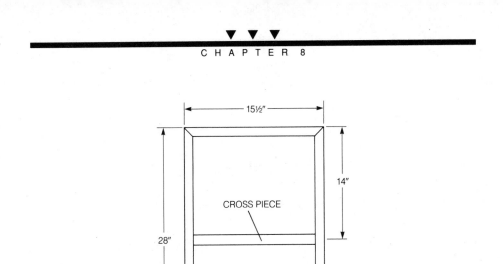

Figure 8-21. Details of the grille frame.

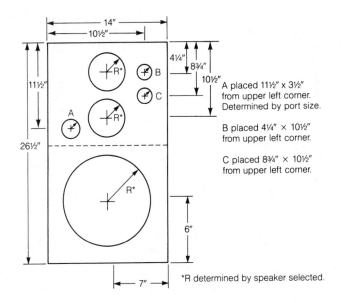

A placed 11½" x 3½" from upper left corner. Determined by port size.

B placed 4¼" × 10½" from upper left corner.

C placed 8¾" × 10½" from upper left corner.

*R determined by speaker selected.

Figure 8-22. Driver board layout for System 4.

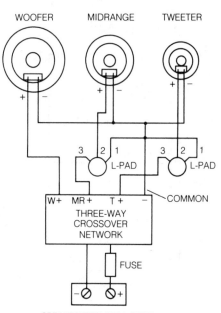

Figure 8-23. Wiring diagram for System 4.

8. Cut four 14-inch and four 26½-inch long battens from scrap pieces of ply-wood. In two 14-inch and two 26½-inch battens, drill, using a #48 bit, a pair of pilot holes, spaced about 1½ inches from the ends. Glue and nail the four battens 3/4 inch from the front of the top, bottom, and sides of the enclosure. The holes you drilled should face the front and back of the box.

9. Place the driver board over the front of the box. Insert No. 6, 1¼-inch wood screws into the previously drilled holes and tighten them into the driver board.

10. Apply caulking to the inside of the enclosure, being sure to seal the cracks between all the joints. Don't get any caulking on the speaker (see *Figure 8-13a*).

11. Mount the other set of four battens 3/4 inch in from the rear opening on the top, bottom, and sides of the box.

12. Cut a hole in the center of the back piece, using the terminal strip as a template. Secure the terminal strip using four 6-32, 1-inch machine bolts and nuts.

13. Staple 1-inch acoustic fiberglass damping to the sides and bottom of the box. Staple a single sheet to the inside of the back panel (see *Figure 8-13b*).

14. Place the back panel inside the box. The back should just barely fit into the box and should be flush with the top, bottom, and sides. Secure the back panel with eight No. 6, 1½-inch wood screws. Apply a light bead of wood glue around the outside joints in the back to provide a good seal.

Finishing the Speaker

Finish the speaker following the procedure given for System 2. When you are finished, apply hook and loop strips to the inside corners of the speaker grille frame and along the outside corners of the enclosure as shown in *Figure 8-24.* Attach the grille to the enclosure.

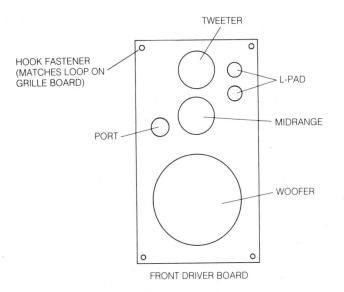

Figure 8-24. Location of hook and loop fasteners.

AUTOMOTIVE
SOUND SYSTEMS

OVERVIEW

The science of sound and the appreciation of music have combined to form an exploding, increasingly sophisticated, audio technology industry. Automobile sound systems have changed dramatically in the past two decades, but still the automobile manufacturers have not kept up with the latest available technology. There are sound systems which, when added to an automobile, can make you feel as if you are at the live performance, and there is no reason why automobile sound systems can't rival, and even surpass, the sound systems intended for use in the home. From small improvements to total system overhauls, this chapter covers ways to customize your automobile listening environment to your own preference.

SUMMARY OF THE AUTOMOBILE SOUND SYSTEM

While factory systems have improved somewhat over the past five years, they have by no means kept up with the times. Since the average person today spends hundreds of hours in an automobile, it is no surprise that people seek a "higher sound" within their automobiles. From compact disc players to equalizers, amplifiers, crossover networks and full-range speakers, many options are available as steps are taken along the road to improve your system.

Characteristics of Factory Installed Speaker Systems

To say the least, today's factory installed automotive sound systems have left something to be desired. Reasons for their inadequacies are limited space, poor speaker placement, undersized magnets and drivers, and sometimes a very limited frequency response.

IMPROVEMENTS USING EXISTING OPENINGS

Significant improvement can be made simply by replacing factory speakers with true high-fidelity speakers. Not only is this the best place to start improving your system, but also the most economical. Some automobile systems may not be making use of all their existing speaker openings; therefore, it is a good

idea to check with the manufacturer for the location and size of the openings in your particular automobile. Added speakers in existing openings can improve a system dramatically.

Figures 9-1 and 9-2 show the common existing speaker openings in an ordinary coupe, sedan or hatchback and a pickup/utility vehicle, respectively. Because pickups have very limited space and have no trunk for the speaker baffle, a box enclosure behind the seat is preferable.

As indicated in Figure 9-3, replacing most factory speakers is usually as simple as removing a cover and four screws, disconnecting two speaker wires from the existing speaker, connecting the wires to the new speaker, and fastening the speaker and cover in place.

NOTE: *Do not assume that the automobile speaker's impedance is the same as that of the speakers in a home sound installation (8 or 16 ohms). The impedance of automobile speakers is often 4 ohms, and even as low as 2 ohms in a few special installations.*

IMPROVEMENTS USING BOX ENCLOSURES

For the real music lover and sound enthusiast, simple after-market speaker replacements probably will not provide sufficient improvement. Box enclosures are often required to obtain the desired full bass sounds which the original system doesn't even reproduce. Through the original system they remain unheard sounds. Many varieties of box enclosures are available. They can be placed in the trunk, under a seat, or behind a seat. These enclosures serve specific functions which were discussed in Chapter 3.

SIGNIFICANT AFTER-MARKET IMPROVEMENTS

There are other ways to improve your existing system in addition to upgrading speakers and their enclosures. Equalizers can be added for more power and definition of sound, faders can be added for balance, and crossover networks can be added for frequency spectrum division to appropriate speakers in multiple speaker systems. High-power amplifiers can be added to gain the forceful sound and the thunderous bass which some people prefer.

Equalizer

Equalizers offer a lot of potential for improving the final sound production. With an equalizer you can fine-tune your music to your individual preference and also compensate for limitations in your vehicle's acoustics and other portions of your system. For example, if your midrange speaker is more efficient than the woofer and the tweeter, you may selectively boost the lows and highs without distorting the midrange frequencies. An equalizer will allow you to do this where the factory-installed tone control or bass and treble controls will not.

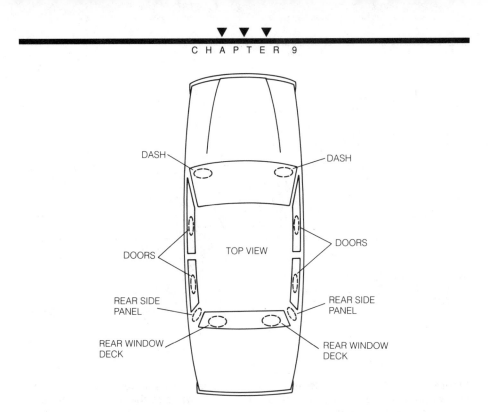

Figure 9-1. The most common existing speaker openings in an ordinary coupe, sedan or hatchback.

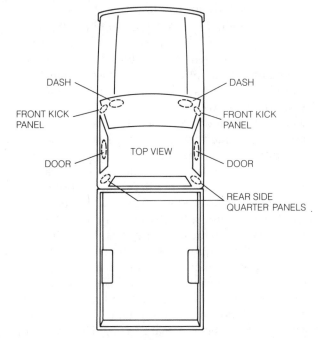

Figure 9-2. The most common existing speaker openings in an ordinary pickup and utility vehicle.

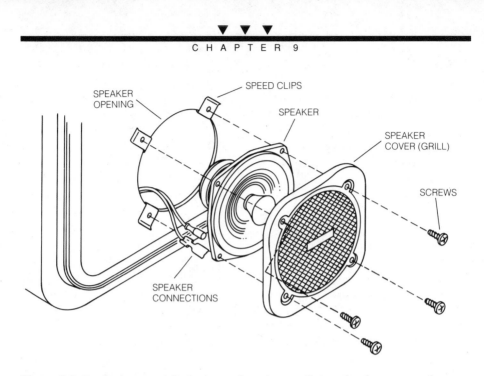

Figure 9-3. Replacing most factory speakers is usually as simple as removing a cover and four screws, reconnecting two speaker wires to the new speakers and reinstalling cover and screws.

An equalizer modifies a signal's frequency spectrum. Essentially it is a filter that can cut or boost specific bands of frequencies. Two types of equalizers are available as shown in *Figure 9-4*. One amplifies the input signal; the other does not. A type that has no amplification is installed as shown in *Figure 9-5* at a point in the system that has line-level in/out jacks. The equalizers that have no amplification introduce some loss in signal level. However, they usually introduce much less distortion (harmonic, intermodulation, and transient) than an equalizer with amplification.

A popular approach today among auto sound buffs is to install an amplifier type equalizer called an equalizer/booster. It includes the equalizer in a power amplifier. An excellent example is shown in *Figure 9-6*. It has seven control bands to allow you to tailor the frequency response in precise 2 dB steps for a total boost/cut range of 12 dB. Frequency response is shown on an 84-LED display, which can be used to see any adjustments which are made in the frequency response. It also splits the 120-watt output: 90 watts go to the rear speakers and 30 watts to the front speakers. The unit shown has both speaker-level and line-level inputs. Some equalizer/boosters have built-in (usually two-way) crossover networks, with either monaural or stereo outputs, to subwoofers. The addition of an equalizer/booster allows more control over the desired sound output from your system and can be used to compensate for weaknesses of other components of the system.

Almost all the equalizers today are analog. However, on the horizon is digital signal processing (DSP), which will allow the equalizers to offer many useful features with extreme high fidelity.

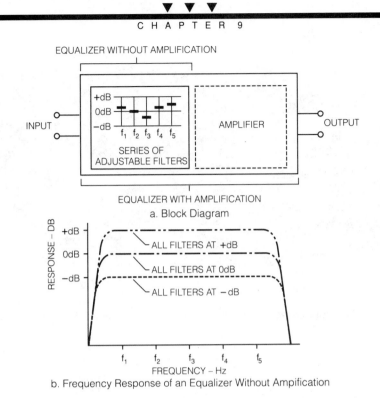

a. Block Diagram

b. Frequency Response of an Equalizer Without Ampification

Figure 9-4. Types of equalizers.

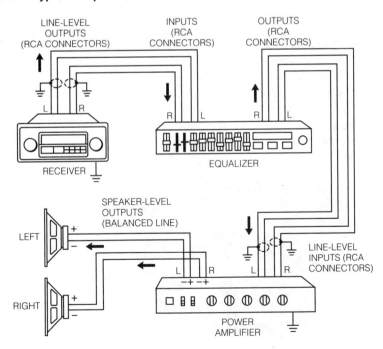

Figure 9-5. Installation connections of an equalizer like that in *Figure 9-4*.

Figure 9-6. An example of a seven-band equalizer/booster. *(Courtesy of Radio Shack)*

Fader

A fader can play an important role in helping to get a "surround sound" effect. A fader is used in a speaker system consisting of four or more speakers. It is used to control the distribution of sound power within the speaker system. Stated simply, the fader adjusts the balance of sound power between front speakers and rear speakers to give you a desired sound. The fader allows you to adjust from total front sound to total rear sound, or a mixed version of each.

Be careful that you do not overpower the fader you choose. *Figure 9-7* shows a fader that will handle a total of 50 watts. Most equalizers have a built-in fader; if yours does, it is not necessary to install an external one. Most faders are installed by simply connecting the speaker leads from your radio to the fader input, and the fader output to the four sets of speaker leads, as shown in *Figure 9-8.* The speaker leads are balanced lines floating from chassis ground and should not be grounded.

Figure 9-7. An example of a fader that will handle a total of 50 watts. *(Courtesy of Radio Shack)*

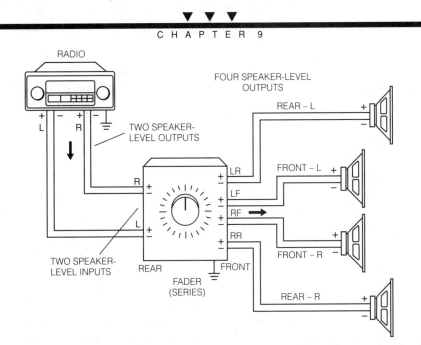

Figure 9-8. Connecting a fader from the output of your radio to the speaker leads.

Crossover Networks

As discussed in Chapter 2, single speakers, which are the most common type installed in factory automobiles, have difficulty in evenly and accurately reproducing all frequencies of the audio spectrum at the same level. Therefore, separate speakers called woofers, midrange, and tweeters were designed to handle particular portions of the frequency spectrum. As discussed in Chapter 6, by routing the various frequencies to a proper driver, you can more fully appreciate the acoustic advantages of reproducing signals with "extended highs and lows" with greater fidelity than any single speaker system can deliver. Crossover networks are used to send signals to the speakers that are best suited for reproducing them.

CAUTION: Sending frequencies to the wrong speakers will not only result in poor sound, but can destroy the speaker driver. Midrange and bass signals should never be sent to a tweeter — not even upper-midrange ones. The tweeter's tiny cone and voice coil cannot withstand the powerful movements associated with the midrange and bass frequencies.

Theory of Operation

Crossovers come in two basic types: passive and active. A passive crossover uses capacitors and inductors to filter out certain frequencies while allowing other frequencies to pass. Capacitors allow high frequencies to pass but attenuate low frequencies. Inductors pass low frequencies but attenuate high frequencies. Chapter 6 discussed two-way and three-way passive crossovers. Because a passive crossover is installed in-line between the power amplifier and the speakers, it will cause a power loss of up to fifteen percent (−0.7 dB). The output power amplifier must be able to supply enough power to overcome this loss.

An active crossover, on the other hand, is installed between the receiver and the power amplifier on the signal-level line. Because the frequencies are split before they are amplified, an active crossover is much more efficient and minimal power loss is incurred.

Power Amplifier

Reason for Inclusion

Most factory car radios are capable of only a few watts output at most. Increasing the available output power not only will help overcome background noises (especially at highway speeds and with the window open), but also will preserve the dynamic variances from soft to loud that give music its "live" sound.

In Chapter 1, we discussed sound levels and illustrated various sound levels in *Figure 1-2*. On the average, a car speaker can produce a sound-pressure level (SPL) of about 88 dB when it is driven by 1 watt of power. Each additional stereo channel increases the SPL by 3 dB. To increase the level by 10 dB requires a ten-times increase in the power, so 10 watts produces 98 dB and 100 watts produces 108 dB per channel. An unamplified jazz band in a small nightclub, or a symphony orchestra in a performance hall, generate live sound levels from a low of about 30 dB up to about 120 dB SPL at loud peaks.

Heavy metal rock groups can reach 130 dB or more. Power amplifiers must be specifically designed to handle such power levels with a minimum of added distortion and noise. An add-on power amplifier especially suited for automotive applications is shown in *Figure 9-9*. It is designed to produce 60 watts per stereo channel.

In a car, the music usually does not sound as loud as it really is. This is because the brain desensitizes the ear in the high ambient sound level environment, especially at highway speeds. This produces a temptation to play the car sound system at louder-than-life volumes which, for some passengers, can cause hearing discomforts.

Installation and Precautions

Here are a number of precautions to remember when installing an amplifier to ensure proper operation and to prevent future problems:

1. The power lead needs to be routed through the firewall and connected directly to the positive terminal of the battery. Use a power tap or solder a lug to the end of the wire. Clean the battery terminal and cable clamp well before making any connection. Loosen the battery terminal cable clamp, slip the solder lug under the cable clamp bolt, and tighten the clamp securely. Cover the connection with grease to keep it from corroding. If an extra length of wire is required, it is very important that an equal or greater sized wire than originally supplied with the amplifier be used for the extension. Insufficient power from the battery due to a voltage drop in the supply leads is often a frustrating problem that plagues a "do-it-yourself" stereo system installation.

Figure 9-9. An add-on 120-watt stereo power amplifier. *(Courtesy of Sanyo Consumer Electronics)*

2. To avoid overheating, do not mount the amplifier in a totally non-ventilated area. Also, when drilling and installing any piece of equipment, be careful not to puncture existing wiring harnesses, automobile components, gas and fluid reservoir tanks, etc.

3. Today's automobiles are constructed with vast amounts of fiberglass, plastic, and other materials that do not conduct electricity. In many respects, this is good, but it does present the problem of finding a good ground connection for the sound system. The important thing to remember is that the sound system ground needs to be connected through a very low resistance conductor to the vehicle chassis. A 12-gauge braided pure copper cable is ideal for this application.

EXTENDED RANGE SYSTEMS

Many people do not realize the importance of the speakers in their sound system. The speaker system is the final step in the sound reproduction system. Its performance specifications must be at least equal to that of the other components in the system, otherwise, it will be the limiting factor in the system. Woofers, midrange, and tweeters are combined to make up quality extended-range systems for a particular application.

Woofers

Recall that woofers are the speakers that handle the low frequencies. With available woofers, you can truly experience low-frequency music reproduction in your car, pickup or van like that in a studio monitor. Using a subwoofer like the 120-watt, 12-inch unit shown in *Figure 9-10,* you can have a widened, flat response to the lowest octave of recorded music. When mounted in a proper

Figure 9-10. A 12″ subwoofer with a dual-wound voice coil and a strong 19-ounce magnet. *(Courtesy of Radio Shack)*

enclosure, it is capable of a tremendous amount of clean, tight bass without power compression. Adding a subwoofer like this to your system will fill out the bottom end of your music and give you the kind of bass that you both hear and feel.

Probably the most important consideration for building a subwoofer speaker system is space. The trunk area in many cars makes the ideal subwoofer enclosure location. Because the lower the frequency the less directional the sound waves, the bass sound from the subwoofers disperses freely and does not need to be aimed directly at the listening area.

If you plan to construct your own enclosure, be sure to review the sections in Chapter 3 on "Estimating Enclosure Volume" and "Calculating Enclosure Size." It is essential that the dimensions include enough air space for the subwoofer to respond properly. Space may limit the use of the golden ratio discussed in Chapter 3, but the enclosure should provide the required total volume for a particular speaker. When the subwoofer is mounted in a subwoofer enclosure like the ones discussed in Chapter 10, and combined with an amplifier with plenty of driving power and a good crossover network, you have a bass system that sounds great and looks good as well.

Midrange

The widest band of frequencies must be handled by the midrange speakers. So much emphasis is placed on the performance at the extreme ends of the frequency response characteristics (the subwoofer and the tweeter) that sometimes the midrange is slighted. This can result in "holes" in the frequency response and "dead sound" areas in any listening environment, but especially in an automobile. The midrange speakers should be capable of high output over a frequency range that overlaps the inside ends of the response of both the woofers and tweeters.

Tweeters

A remarkably noticeable improvement can be made in the treble clarity of your system by adding tweeters. They add clean and crisp highs, especially in female vocals, cymbals and snare drums, by extending the speaker system's flat frequency response well past the range of human audibility. To reproduce this upper audible range, a tweeter must move up to more than 25,000 (25 kHz) complete vibrations per second. This rapid change in motion requires the lightest possible cone or radiating surface, but it also requires a material that can withstand the tremendous stress of an acceleration of up to 1000g without any deformation. The tweeter should reproduce the high frequency electrical signals with the most efficient and true transfer of energy. It should also have fast transient response with minimum coloration. This is accomplished in the 3¾ inch flared-horn tweeter shown in *Figure 9-11* by using a piezoelectric ceramic element.

Figure 9-11. A 3¾" flared-horn tweeter using a piezoelectric ceramic element.
(Courtesy of Radio Shack)

Unlike the woofers, the tweeters are mostly self-contained. They do not need an acoustic enclosure, but the mounting is critical. They should be mounted high (preferably in a hole) and in a direct, unobstructed line with the listening area to avoid what is called a "dead sound."

Full-Range Speakers

Extended range systems are fine, but if one chooses to just replace a speaker in an existing hole, is there a way to provide any of the extended range sound? Yes, by using a high-performance, full-range speaker. The full-range speakers are made of woofers and tweeters and/or midrange drivers all in one assembly that cover the full audio frequency spectrum. This type of replacement should be of higher power rating than the factory installed speakers to make use of all of the audio power that the radio will deliver without overdriving the speakers. In *Figure 9-12*, three types of full-range speakers are shown for mounting in the door, rear deck and dash (be sure to check your existing hole sizes).

a. Door Mounting
(Courtesy of Sanyo Consumer Products)

b. Dash Mounting
(Courtesy of Radio Shack)

c. Rear-Deck Mounting *(Courtesy of Radio Shack)*

Figure 9-12. Three types of full-range speakers.

If you do not wish to use the factory holes or cut new ones, surface-mount speakers may be your answer. Several examples are shown in *Figure 9-13*. Such speakers have the added advantage of already having their enclosure. This means a minimum effort to mount the speakers. The very high quality 4-way shown in *Figure 9-13a* includes a bass-reflex designed enclosure. *Figure 9-13b* shows a mini 2-way surface-mount unit for tight installations like that found in subcompact automobiles.

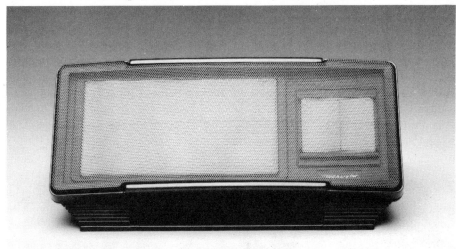

a. High-Quality 4-Way *(Courtesy of Radio Shack)*

b. Mini 2-Way *(Courtesy of Sanyo Consumer Products)*

Figure 9-13. Examples of surface-mount speakers.

In the next chapter, we will cover specific automotive installations. Since automotive vehicles are quite varied in size and shape, it is rather difficult to provide plans for a universal sound system; therefore, suggested installations are provided for coupes/sedans, hatchbacks, pickup trucks and sport utility vehicles as examples of what can be done.

CHAPTER 10

SPECIFIC AUTOMOTIVE INSTALLATIONS

Although there are many styles of automobiles today, there are different ways to tailor the sound system to each. Some of the possibilities to consider for the speaker system upgrades are: full-range multiple-driver box enclosures, subwoofer box enclosures, and the use of existing trunk, door or factory openings or enclosures.

It would be virtually impossible to cover installations for every type of car on the road today, so we will consider specific speaker installations in these primary vehicle styles:

1. The coupe/sedan (two door and four door with a trunk)
2. The hatchback (two door with no trunk)
3. The pickup truck (with its limited space)
4. The large, boxy van and sport utility vehicles

Even if your exact vehicle is not covered, you will be able to determine which general style it falls in. If you apply the installation procedures recommended in this chapter and the construction techniques outlined earlier, you can create a quality system that sounds great.

COUPE/SEDAN

Sedans and coupes have some helpful features when it comes to stereo modification. The trunk is an excellent location for a boxless enclosure because it provides an effective speaker baffle. Also, when mounted in the trunk, speakers and amplifiers are hidden from the occupied area of the vehicle. Most coupes and sedans built today have factory speaker openings that are properly placed and sized for better than average sound.

Improvements in sedans and coupes can be approached in phases. Depending on your particular stock stereo system, you can gain a considerable improvement in sound by simply locating existing speakers/openings, determining their sizes and replacing the speakers with a good-quality, full-range, after-market speaker as discussed in Chapter 9.

If you want further improvements, the next step we recommend is replacing the stock receiver with a high-quality, after-market, AM/FM stereo receiver. You may want one that has an audio tape cassette player or a compact disc player. Connect it to the newly installed speakers and you should enjoy a much improved sound system. An example AM/FM stereo cassette and separate compact disk player are shown in *Figure 10-1*. Because of the relatively poor fidelity, high distortion, and low output power of factory installed radios, you probably will be surprised by the improvement resulting from the simple changeover from the stock equipment to an after-market speaker and radio.

If your taste craves still more, the next step would be to increase the system's audio power. Read the specifications on the speakers you have installed and obtain an amplifier with a power output rating close to the speakers' maximum power rating, but don't exceed the speakers' ratings. For example, if the speakers are rated for a maximum of 75 watts each, then an amplifier with a rating of at least 60 watts per channel, but not more than 75 watts per channel, should perform the job very well.

a. AM/FM Stereo Cassette
(Courtesy of Sherwood/Inkel, USA)

b. Compact Disc Player
(Courtesy of Sansui)

Figure 10-1. AM/FM stereo cassette receiver that replaces factory-installed receiver and a compact disk player for a system add-on.

The same method of power selection can be used if an equalizer/power booster is what you want. While equalizers are usually mounted in or under the dash for ease of adjustment, amplifiers usually can be mounted in the trunk because repeated control adjustments on them are not necessary. The manufacturer's instructions included with the amplifier usually describes detail methods for mounting and wiring each particular unit.

CAUTION: ANY ADDED COMPONENT, SUCH AS SPEAKERS, ENCLOSURES, EQUALIZERS, AMPLIFIERS, ETC., INSTALLED IN AN AUTOMOTIVE VEHICLE, MUST BE SECURELY ATTACHED IN PLACE TO PREVENT PERSONAL INJURIES IN THE EVENT THE VEHICLE IS ENVOLVED IN AN ACCIDENT.

Routing the wires from the front to the rear of the automobile is sometimes a troublesome task. One of the best ways is to remove the door threshold screws and the threshold itself as shown in *Figure 10-2*. Next, gather all the wires intended for this path and tape them together every 8 to 10 inches to keep them neat and easier to handle. Finally, lay the wires in the door groove and re-install the threshold. Be careful not to damage any of the wires by screwing into them or laying them where the door will pinch them. If the wires are too bulky for one groove of the threshold, separate them and tape them in separate grooves.

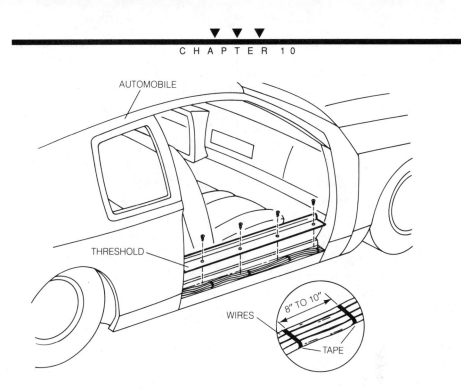

Figure 10-2. Routing wires under threshold plate from front to rear of an automobile.

Because the trunk offers an excellent woofer enclosure, quite a bit of bass can be obtained by simply installing a pair of high quality woofers on the rear deck. However, full-range speakers may be necessary due to limited mid-range and tweeter mounting space. Many of today's cars offer a pair of speaker openings in the rear deck which will accommodate 6″ × 9″ speakers. If this is the case, and if full-range sound is required, a good replacement would be a high-capacity three-way speaker as shown in *Figure 10-3*. This particular speaker offers full-range sound: bass frequencies from its 6″ × 9″ polypropylene woofer with 16-ounce magnet, mid-range frequencies from its 2½″ mid-range speaker, and high frequencies from its 1½″ piezoelectric tweeter.

Even though the trunk-equipped automobiles offer some of the best bass sounds without constructing some kind of after-market box enclosure, many people strive for a deeper, thunderous sound. As seen in the coupe style 1984 Chevrolet Monte Carlo® in *Figure 10-4*, a sub-woofer box can be bought or built to go in the trunk. The box needs to be positioned against the back seat of the car so the woofers are firing directly through the back seat. Obtain accurate measurements for the available space. It would be very frustrating to build an enclosure and find that it doesn't fit when you try to install it.

The low frequency bass tones will have no trouble passing through the rear seat. However, there is no need to install mid-range or tweeters in a trunk-mounted box since the higher frequencies will not be heard through the seat. You can rely on the front speakers for the middle and high frequencies. We recommend that you use at least an 8″ subwoofer, like the one shown in *Figure 10-5*, in a trunk-mounted box. You might use a larger one if space allows.

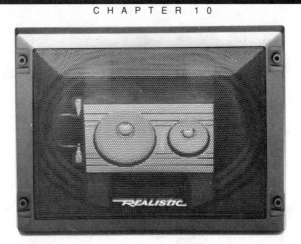

Figure 10-3. Replacement full-range 6″× 9″ speaker. *(Courtesy of Radio Shack)*

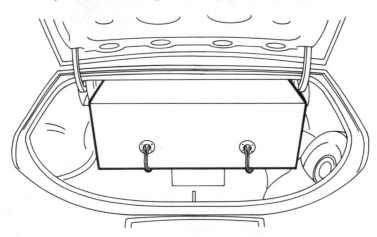

Figure 10-4. Box enclosure installed in the trunk of a coupe-style 1984 model Monte Carlo.

Figure10-5. Subwoofer (8″) for use in trunk-mounted enclosure of *Figure 10-4.*
(Courtesy of Radio Shack)

We strongly recommend that, when possible, a speaker grille or cover be used to protect the speaker from damage. It also limits speaker muffling which impairs cone throw. An example of an 8-inch wire mesh grille that can be applied to the above referenced woofer is shown in *Figure 10-6*. Speaker grilles are usually installed by simply snapping them to the front of the speaker, or by securing them with wood or metal screws, depending on your particular application.

Figure 10-6. 8″ wire mesh speaker grille.

When building your box, be sure that you have enough air volume for the chosen speakers (refer to Chapter 3). Limited or excessive air space causes improper speaker cone throw which results in untrue sound, distortion and even permanent damage to your speakers.

Figure 10-7 shows an example of a trunk-mounted box which contains two of the 8″ subwoofers shown in *Figure 10-5*. It is made from 3/4″ particle board. Refer to Chapter 4 for some specific construction techniques. After completing construction, the box can be positioned and anchored in place with brackets as shown in *Figure 10-7*. Handles are provided on the side of the enclosure for easy handling.

HATCHBACK

The hatchback is one of the newer automobile styles on the road today, with only a few versions going back before 1982. While they share some similarities with the coupe-style automobile, they are different in many ways.

To obtain a good true bass, a speaker enclosure is usually required because there is no trunk for a natural baffle. Factory speaker holes can range from the dash to the doors to the rear side panels, or combinations of all three. As in the coupe or sedan, we recommend that you replace these with full-range, high quality, after-market speakers as the first step.

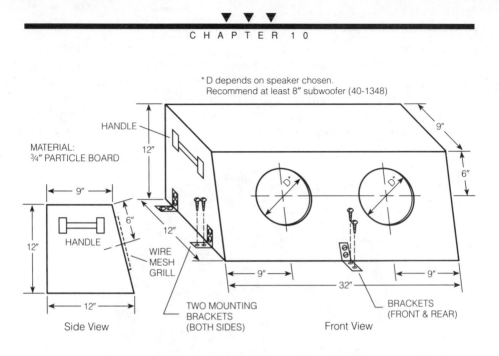

Figure 10-7. Example of basic trunk-mounted enclosure with two 8″ subwoofers.

Since a separate enclosure with separate amplifier is required, a more sophisticated system using multiple amplifiers should be considered. When designing multi-amplifier systems (two or more), it is important to balance the power and send the frequencies to the appropriate speakers. This can be done easily and efficiently by using an equalizer like the one shown in *Figure 10-8* This particular equalizer has no power gain (amplification) and has a built-in, selectable subwoofer crossover. When this equalizer is installed in a system as shown in *Figure 10-9*, it provides a control center that enables you to adjust the front-to-rear balance, adjust tone qualities, and permits a true subwoofer system which can be switched in below the 60 Hz or 100 Hz crossover points.

Figure 10-8. Equalizer with built-in selectable subwoofer crossover.
(Courtesy of Sanyo Consumer Electronics)

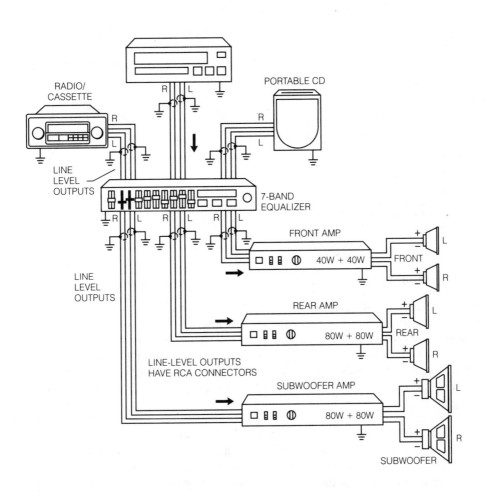

Figure 10-9. Complete automobile sound system with equalizer feeding booster amplifiers for front and rear speakers in existing mountings and a separate sub-woofer amplifier for enhanced bass.

If you chose to build or purchase a box enclosure, you have several choices. A box consisting of woofers, midrange speakers and tweeters will work superbly. However, even though they will add somewhat to the final sound, the midrange speakers and tweeters are not necessary. As in a coupe, the mid- and high-frequency sounds are usually handled nicely by the high-quality full-range speakers that replace the factory speakers. Reproduction of these sounds in the rear box may never be heard.

A good example of a box that will fit in many of the hatchbacks on the streets today is shown in *Figure 10-10.* Once again, be sure to measure the amount of available mounting space in your particular hatchback. Be sure that your enclosure is not too tall for the rear hatchback door to close freely, completely avoiding contact with the box.

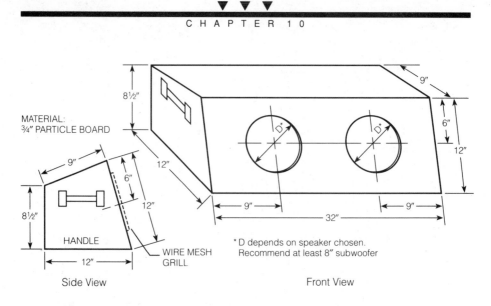

MATERIAL:
¾" PARTICLE BOARD

8½"

9"

6"

12"

8½"

HANDLE

12"

WIRE MESH GRILL

Side View

9"

12"

12"

9"

9"

32"

9"

6"

12"

* D depends on speaker chosen.
Recommend at least 8" subwoofer

Front View

Figure 10-10. Example of enclosure that will fit in many hatchbacks.

After completing the construction of the box, simply mount it and connect wires as described previously. The low-frequency sounds that woofers reproduce are not as directional as mid-range and higher sounds. Therefore, we advise that the woofer enclosure be mounted as far to the rear of the hatchback area as possible. This allows a fuller, more surrounding sound and tends to keep the bass sounds from overpowering and possibly muffling the higher frequency sounds.

Because the speaker enclosure that is installed in a hatchback is visible to the occupants, we suggest that you either paint, stain, or cover the enclosure with carpet or fabric to make it more appealing to the eye. Usually carpet can be purchased from the automobile manufacturer that will match the factory carpet installed in your car. To install carpet on a wooden box, first cut the carpet to the dimensions of the exterior panels of the enclosure. Next, apply carpet adhesive to the box (according to manufacturer's instructions), then stretch and smooth the carpet in place and allow to dry. Carpet tacks or staples may be placed around the edges of the carpet for extra binding.

PICKUP TRUCKS

Pickup trucks are no longer used solely as farm and construction vehicles. With their continual rise in popularity, owners seek ways of improving their sound system. Of the four basic styles of automobiles we are covering, the pickup truck is probably the most challenging to enhance.

Because of very limited space, it is difficult to install a proper speaker enclosure. While high and mid-range sounds pose some problems, they are minor compared to the problem of acquiring the good, tight low-end bass sought by most music enthusiasts. Because of the physical size of the woofer speaker necessary to reproduce the bass sounds and the limited space, some impairment of the final sound is hard to avoid.

As if speaker location were not problem enough, amplifier installation space is close to none. A very small amplifier (e.g., 3″× 5″× 2″) can be mounted under the dash, but anything larger must be installed under the seat. Even this presents a clearance problem. Since most pickups have an adjustable seat, plan on installing an amplifier that allows free movement of the seat.

Many people tend to think removal of a seat is far more difficult than struggling to work under the seat. More often than not, this is untrue. By simply removing four bolts, in most cases, and enlisting the assistance of a friend to lift the seat out of the vehicle, you can make your job considerably easier and less frustrating.

The seats in most automobiles are firmly mounted to the body floor. Simply connecting the grounding wire from the amplifier to the seat bolt provides an excellent electrical ground. Also be careful while mounting the amplifier to the floor under the seat to avoid drilling and/or screwing into existing wiring. Solid mounting is strongly recommended so the amplifiers will not move while traveling over bumpy roads. If they move around, the power and speaker wires may become disconnected or shorted.

Since the only space available for speaker mounting is behind the seat, it can be very difficult to obtain a pleasing surround-sound effect. Factory speaker openings on many pickups are limited to the dash, door, or front quarter panel, so it takes quite a bit of inventive work to obtain satisfactory performance. If mid-grade improvement is your goal, then we recommend upgrading the factory dash or door mounted speakers first, and then upgrade the AM/FM stereo equipment, perhaps adding a cassette or compact disk player. Next, installing a pair of high power 6″× 9″ full-range speakers behind the pickup seat may provide the sound you want. These can be installed in simple enclosures like the one shown in *Figure 10-11*. Some pre-fabricated ones maybe available in the marketplace.

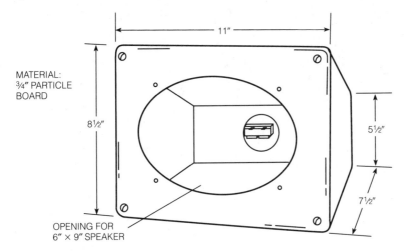

Figure 10-11. Basic enclosure for 6″× 9″ speakers.

If after you try this you still want louder and more tailored sound, we recommend you use an equalizer/booster such as the one shown in *Figure 10-12*. This model offers variable volume control, CD input, defeat, front-to-rear four-speaker fader, and 70 watts of power.

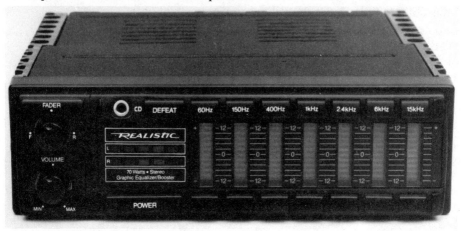

Figure 10-12. 70-watt equalizer/booster. *(Courtesy of Radio Shack)*

Figure 10-13 shows a scaled drawing of a 1984 Chevrolet Silverado® pickup. Notice that it takes every inch of space to enclose the two massive 12-inch woofer speakers. An excellent way to help establish a full-range sound in a pickup is to install a pair of tweeters in the top of the box firing towards the ceiling. This is shown in *Figure 10-14*, which is an enclosure that may fit behind

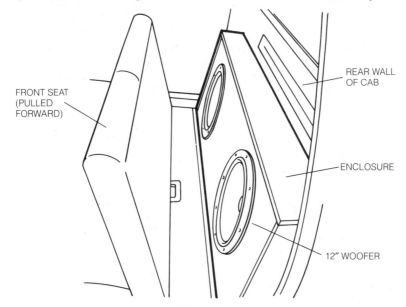

Figure 10-13. Two 12″ woofers installed behind the seat of a 1984 Chevrolet pickup.

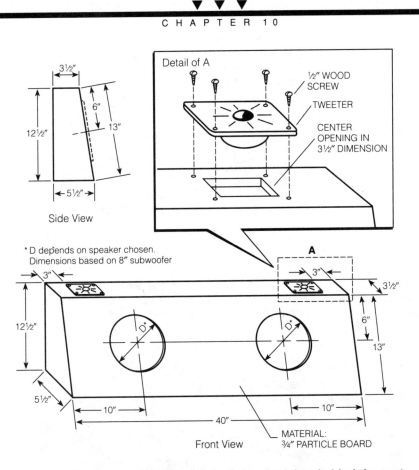

Figure 10-14. Tweeter positioning and mounting in pickup behind-the-seat enclosure.

the seat of several style pickups. It is designed for 8″ sub-woofers. The tweeter balances out the bass tones reproduced by the woofers. Remember to measure the space available before beginning to build any enclosure. Also, apply the sealing and finishing techniques of Chapter 5 as required.

Figure 10-15 shows that the factory radio has been removed and an AM/FM-compact disc player installed in its place. Also a "slim line" equalizer with built in subwoofer crossover is mounted under the dash.

If the design and construction of a box is not feasible, today's market offers a wide selection of special pickup style boxes. *Figure 10-16* shows a good example of a pair of pickup boxes with 8″ woofers and top firing piezoelectric tweeters. Because of their compact design (12.5″× 18.5″× 6.25″), they will fit behind the seat of most pickups. Proper mounting of these ready-to-install boxes can be done by temporarily removing the woofer drivers from the boxes, setting the boxes in place, and screwing a "self-drilling" screw through the back of the box directly into the rear wall of the truck. Repeat this procedure three or four times for each box to ensure stability.

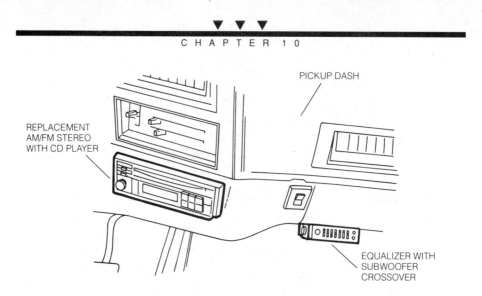

PICKUP DASH

REPLACEMENT
AM/FM STEREO
WITH CD PLAYER

EQUALIZER WITH
SUBWOOFER
CROSSOVER

Figure 10-15. Upgrade system in 1984 chevrolet pickup where the factory radio was replaced with an AM/FM stereo with CD player. Equalizer with no amplification but with subwoofer crossover was mounted under the dash.

Figure 10-16. Pickup boxes with 8″ woofers and top-firing piezoelectric tweeters.
(Courtesy of Radio Shack)

All in all, an excellent sounding system can be obtained in pickup trucks. However, it is important that the dimensions of space available in your particular vehicle be measured and noted before beginning the system enhancement project.

VANS AND SPORT UTILITY VEHICLES

Vans and sport vehicles are the largest of the vehicle styles that we will discuss. The large interior space presents a different kind of problem; that is, there is a large volume of space to be filled with sound. This may mean that higher wattage amplifiers and speakers are needed. On the other hand, there is plenty of room to install the hardware.

Vans

In both full-size and mini-vans, factory radios are easily replaced with after-market versions in a wide choice of compatible units. Mini-vans have factory openings in the dash, doors and rear quarter panels. After-market replacements in existing openings can provide significant improvements. Full-sized vans, on the other hand, usually have factory openings that are limited to the dash and door. As a result, box enclosures are usually required for improved sound.

One might choose just to build an enclosure with a woofer. It would certainly make an improvement, but if a quality enhanced system is desired, a full-range box is recommended. The box should contain adequately sized woofers—a minimum of 8 inches, a good quality midrange speaker firing towards the front of the van, and a suitable horn or dome tweeter mounted so there are minimal obstructions and aimed toward the center listening area. Of course, prefabricated boxes can be used, but here is an opportunity to apply the design techniques for home systems to an automotive system—and save some money at the same time.

If you choose to construct the box yourself, it should be made of 3/4″ particle board. An example of an average box containing two 12″ woofers, two 4″ midrange speakers, and two 4½″ horn tweeters is shown in *Figure 10-17*.

Figure 10-17. Typical box enclosure containing two 12″ woofers, two 4″ midrange speakers, and two 4½″ horn tweeters. The midrange speakers and tweeters fire vertically. *(Courtesy of Sound Works)*

Enclosures for such applications are designed using a 12" subwoofer for the woofer, a 5¼" midrange speaker, and a 3¾" flared-horn piezoelectric tweeter. If a rear bench seat exists, this is a prime spot for speaker and amplifier mounting, so the enclosure to be built can be adjusted to fit in a particular space.

The final size of the holes that need to be cut for the speakers is determined by referring to the speaker specifications that come with the speaker. *Figure 10-18* is sort of a universal size enclosure that should house any of the previously mentioned speakers. It is designed to sit on the floor in the back of a full-sized van. It has the following dimensions: width 36", height 14½", depth 12".

After completing construction of the box, mounting the speakers and crossover networks, wiring the speakers, and finishing the enclosure, simply position the box in the rear of the van. Many enclosures like this for vans are covered with carpeting to match the van's interior. Mounting the box directly to the van can be accomplished by either screwing directly through the wood of the box into the van chassis, or by placing brackets, similar to those shown on *Figure 10-7*, on the sides of the box and affixing them to the chassis.

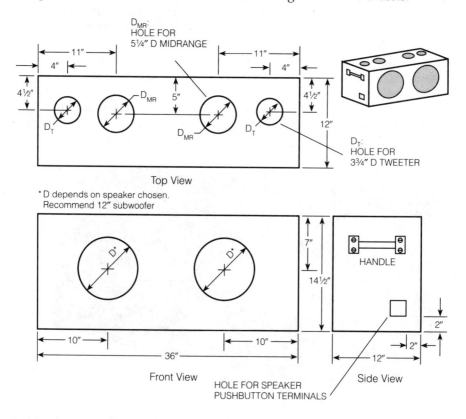

Figure 10-18. Standard size box enclosure that works well in a van.

Look for a suitable space to install the amplifier(s). A common place for the amplifier(s) is under the rear seat, next to the box. If this is not feasible, special shelves or ventilated boxes can be mounted to the interior walls to hold the amplifiers, or they may be attached directly to the box enclosure itself. It is best to choose a secure position so that the amplifier controls can be adjusted, but are not available to everyone.

Sport Utility Vehicles

Basically the same procedures covered for van applications can be applied to the popular utility sports vehicles (e.g., Chevrolet Blazer® and Ford Bronco®). *Figure 10-19* shows the enclosure of *Figure 10-18* installed in the rear "cargo bay area" of a 1989 Chevy S-10® Blazer. Remember, the larger the speakers, the more power needed to drive them efficiently.

Figure 10-19. Standard size box enclosure installed in back of a 1989 model S-10 blazer. Speaker terminals are on rear of this enclosure. *(Courtesy of Sound Works)*

To complete this speaker system, you would probably want to replace the existing factory speakers with high-quality after-market speakers. This particular automobile has two 4" × 6" factory openings in the dash and two 6" × 9" factory openings in the rear quarter panels.

All in all, depending on the caliber of system you want to invest in, astounding stereo systems can be tailored into today's vans and utility vehicles.

In the world today change is everywhere. Everything old is challenged and often redefined to new standards and new expectations. Modern audio technology and the appreciation of music have combined to provide outstanding audio systems for the automobile, to the point of forming a whole new after-market industry. The wait for high-fidelity stereo performance in the automobile, no matter what the type, is now over. If you have set a goal to improve your system, we hope that the ideas and concepts in this book, with the emphasis in the last two chapters on automotive systems, will help you achieve this goal. If it does, then we have met our goal.

Design Equations

NOTE:

The design equations for speaker systems are very complex, and have many variations based on the design requirements. The equations in this appendix were chosen to give the best system performance for systems used in the home. For example, the base reflex systems will have a flat response and highest efficiency for a particular Q. The equations also were selected to provide easy "rules-of thumb" for the hobbyist who wants to build a single-woofer system for the home with readily available components. They track best with actual performance when speakers are used that have Qts between 0.3 and 0.5 and Vas from 6 to 12. For professional applications, like high-powered sound reinforcement for auditoriums, a different set of design criteria are required and different equations must be used.

The easy-to-follow design charts provided in Chapter 3 allow you to construct a good sounding speaker system using almost any woofer. The charts assume a woofer with average characteristics,and do not take into consideration wide variances in Vas, Qts, and free air resonance specifications. The equations in this appendix provide a means to more accurately calculate speaker enclosure parameters given a specific woofer. Included in this appendix are:

- Design equations for determining the ideal volume for acoustic suspension and most ported reflex speaker enclosures. Instructions are given for using a basic four-function calculator to enter values in the equations.
- Design equation for computing the resonance frequency of a woofer installed in a box of a specific size. Instructions are given for using a basic four-function calculator to enter values in the formula.
- Design equations for calculating optimum enclosure volume and tuning for ported reflex systems. A scientific calculator is needed for these equations.

ACOUSTIC SUSPENSION ENCLOSURE VOLUME

Use these equations for computing the optimum volume for anacoustic suspension speaker enclosure. These equations will also work with most woofers placed in a ported reflex enclosure. Before doing the calculations, determine these specifications for the woofer you wish to use:

- Speaker compliance, rated in Vas (convert to cubic feet if necessary)
- Q of the speaker, usually listed as Qts
- Free-air resonance, in Hz, typically stated as Fo or Fs

Calculations

1. Decide on an overall Q for the speaker enclosure using the chart in *Figure A-1*. A Qb of 0.7 provides the flattest overall frequency response with no peaking. Decreasing enclosure volume will increase system Q, depressing the very deep bass tones while emphasizing slightly higher frequencies. A Qb of 1.0 represents a good compromise between deep bass and flatness of upper bass response. Qb values outside the range of 0.5 to 1.5 are not recommended.

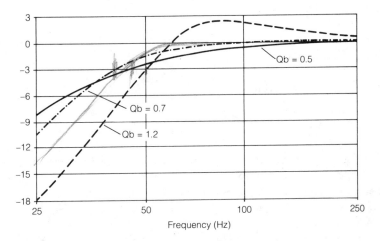

Figure A-1. Effect of Qb on Frequency Response for a 12-inch Speaker with Fo = 25 Hz, Qts = 0.4, and Vas = 10 cubic feet

2. Find the ratio (Qr) between box Q (Qb) and speaker Q (Qts) with the equations:

$$Qr = Qb/Qts$$

3. Calculate the volume ratio (Vr) with the equation:

$$Vr = Qr^2 - 1$$

4. Calculate the volume of the box, in cubic feet, with the formula:

$$Vb = Vas/Vr$$

Example

This example shows how you would use these equations with a four-function calculator.

Assume:
      Qts of speaker is 0.51.
      Vas of speaker is 2.50 cubic feet.
      Desired Q (Qb) of finished box is 1.2.

- To find Qr, press these keys:
  1 . 2 ÷ 0 . 5 1 =
  Result: 2.35
  Write down the result for use in the next step.

- To find Vr, press these keys:
  2 . 3 5 × 2 . 3 5 =
  Result: 5.52
  Press: −1 =
  Result: 4.52

Note that the square of Qr (2.35) was calculated by multiplying it by itself.
Write down the result for use in the next step.

- To find Vb, press these keys:
  2 . 5 0 ÷ 4 . 5 2 =
  Result: 0.553

The result of the calculation example is 0.553 cubic feet, or 955.5 cubic inches (convert cubic feet to cubic inches by multiplying by 1728). Note that the actual volume of the enclosure does not need to be precisely 0.553 cubic feet. A variation of as much as 10 percent usually has no noticeable effect on the quality of the sound.

Once you know the required volume, you can calculate the best dimensions using the Golden Ratio described in Chapter 3. You can also refer to the tables in Appendix B.

ACOUSTIC SUSPENSION ENCLOSURE FREQUENCY

Varying the enclosure volume not only affects system Q, but it also changes the in-box speaker resonance, the low frequency cutoff point (where low frequency response is down 3 dB), and the amount of response peaking above resonance. These effects are illustrated in *Figure A-1* and *Table A-1*. To calculate the in-box resonant frequency of the speaker, you need to know the free-air resonance of the speaker and its in-box Q. The in-box resonant frequency can then be used to determine the − 3 dB response frequency with the aid of the system Q multiplier given in *Table A-1*.

Table A-1. The Effect of Qb on Box Volume, Resonance, –3 dB Point and Peaking for the Speaker of Figure A-1 (12")

Qb	Vb (cu. ft.)	Fsb (Hz)	F₃ (Hz)	Peaking (dB)
0.5	17.8	31.3	48.4	none
0.7	4.9	43.8	43.8	none
1.0	1.9	62.5	49.4	1.25
1.2	1.25	75.0	55.5	2.41
1.5	0.77	93.7	65.6	4.03

Table A-2. System Q Versus Enclosure Low Frequency Cutoff

Qb System Q	Qm Low Frequency Cutoff (Response Down 3 dB) as a Ratio of System Resonance Frequency
0.5	1.55
0.6	1.22
0.7	1.00
0.8	0.90
0.9	0.83
1.0	0.79
1.1	0.76
1.2	0.74
1.3	0.72
1.4	0.71
1.5	0.70

Calculations

Follow these steps to calculate the in-box resonant frequency (Fsb) and the low frequency cutoff point (F_3).

1. Use the value of Qr in the box volume calculation to compute the in-box resonant frequency for the speaker as follows:

$$Fsb = Qr \times Fo$$

2. Using your desired value for Qb, look up the system Q multiplier (Qm) in *Table A-2*.
3. Calculate the cutoff frequency of the enclosure with the equation:

$$F_3 = Fsb \times Qm$$

Example

This example shows how to use these equations with a four-function calculator.

Assume:

Free-air speaker resonance of 45 Hz.

- Using the value of Qr from the box volume calculation (2.35), compute Fsb by pressing these keys:

  2 . 3 5 × 4 5 =
  Result: 105.75 Hz

- Find Qm in *Table A-2* for the desired Qb of 1.2.

  Result: 0.74
  Compute F_3 by pressing these keys:
  1 0 5 . 7 5 × . 7 4 =
  Result: 78.26 Hz

The low frequency cutoff of the enclosure, using the speaker Q, and volume indicated in this and the previous section is 78.26 Hz.

PORTED REFLEX SPEAKER EQUATIONS

The design equations for ported reflex enclosures are much more involved and may not work with all speakers. The following equations are based on the so-called Thiele/Small parameters,which are used by many speaker and speaker enclosure makers.

To use the equations, you must use a scientific calculator with a log or exponent key and know these specifications of the woofer you are using:
- Fo — free air resonance of speaker
- Qts — Q of speaker
- Vas — compliance of speaker
- Equation for calculating enclosure volume:

$$Vb = Qts^{2.87} \times 15 \times Vas$$

To use the equation, raise Qts to the 2.87 power and multiply the result by 15. Finally, multiply that result by Vas. .
- Equation for calculating bass cut-off frequency:

$$F_3 = Qts^{-1.4} \times 0.26 \times Fo$$

To use the equation, raise Qts to the –1.4 power and then multiply the result by 0.26. Finally, multiply that result by Fo.

- Equation for calculating frequency of resonance for the enclosure

$$Fb = Qts^{-0.9} \times 0.42 \times Fo$$

To use the equation, raise Qts to the –0.9 power and then multiply the result by 0.42. Finally, multiply that result by Fo. Select the port diameter and duct length from *Table 3-1* or *Table 3-2* to obtain the calculated value of Fb in the box with calculated volume Vb.

Notice that smaller values of Qts result in smaller boxes and higher cut-off frequencies. Conversely, larger values of Qts produce larger boxes and lower cut-off frequencies. The ported reflex design equations are valid for values of Qts less than 0.8; however, for best results, values for Qts of 0.5 or less are recommended.

Finally, an equation for calculating the duct length is given below. The equation may be used in place of *Tables 3-1* and *3-2*. It is written in terms of the duct area, and may, therefore, be used with a duct of square or rectangular cross section, or with circular ducts of diameters different from those given in the tables. The equation for calculating duct length is:

$$Ld = \frac{2691 \times Sd}{Vb \times Fb^2} - 0.88 \times \sqrt{Sd}$$

Where: Ld = duct length in inches
Sd = duct area in square inches
Vb = enclosure volume in cubic feet
Fb = enclosure resonant frequency in Hz

To use this equation, first multiply the duct area you wish to use (Sd) by 2691. Next divide this intermediate result by the volume, Vb, and then divide this second result by the square of Fb (i.e., Fb $\times$ Fb). From this number subtract 0.88 times the square root of the duct area, Sd, to obtain the final answer.

The area of a circular duct opening is given by:

$$Sd = 0.7584 \times D^2$$

where: Sd = duct area in square inches
D = duct inside diameter in inches

The area of a rectangular duct opening is given by:

$$Sd = W \times H$$

where: W = duct inside width in inches
H = duct inside height in inches

Enclosure Dimension Tables

These tables indicate the "ideal" inside dimensions for various sizes of woofers and speaker enclosures. Note that each table is identified by the woofer diameter and the the total number of speakers in the enclosure. The dimensions for height, width, and depth are based on the Golden Ratio. (Refer to Chapter 3 for more information on the Golden Ratio.) The width dimension was used as the base, and the fractional height and depth dimensions were chosen to closely approximate the volume calculated using the Golden Ratio dimensions. Enclosure size was increased slightly for two-way and three-way systems. No dimension deviates,more than about 10 percent from the optimum ratio of:

Width = 1.0
Depth = 0.6 times width
Height = 1.6 times width

Table B-1. 4-inch Woofer — One Way

Volume (cu. in.)	Volume (cu. ft.)	Height (in.)	Width (in.)	Depth (in.)
264	0.15	10⅜	6½	3⅞
329	0.19	11¼	7	4¼
405	0.23	12	7½	4½
492	0.28	12¾	8	4¾

Table B-2. 4-inch Woofer — Two Way

Volume (cu. in.)	Volume (cu. ft.)	Height (in.)	Width (in.)	Depth (in.)
265	0.15	10½	6½	3⅞
331	0.19	11⁵⁄₁₆	7	4¼
407	0.24	12⅛	7½	4½
607	0.35	12⅞	8	4¾

Table B-3. 4-inch Woofer — Three Way

Volume (cu. in.)	Volume (cu. ft.)	Height (in.)	Width (in.)	Depth (in.)
413	0.24	12⅛	7½	4⅝
501	0.29	12⅞	8	4⅞
608	0.35	12¾	8½	5¼

Table B-4. 6-inch Woofer — One Way

Volume (cu. in.)	Volume (cu. ft.)	Height (in.)	Width (in.)	Depth (in.)
492	0.28	12¾	8	4¾
594	0.34	13⅝	8½	5⅛
700	0.41	14⅜	9	5⅜
960	0.56	16	10	6

Table B-5. 6-inch Woofer — Two Way

Volume (cu. in.)	Volume (cu. ft.)	Height (in.)	Width (in.)	Depth (in.)
493	0.29	12⅞	8	4¾
592	0.34	13¹¹⁄₁₆	8½	5⅛
702	0.41	14½	9	5⅜
966	0.56	16⅛	10	6

Table B-6. 6-inch Woofer — Three Way

Volume (cu. in.)	Volume (cu. ft.)	Height (in.)	Width (in.)	Depth (in.)
51	0.03	12⅞	8	4⅞
711	0.41	14½	9	5½
985	0.57	16¼	10	6⅛

Table B-7. 8-inch Woofer — One Way

Volume (cu. in.)	Volume (cu. ft.)	Height (in.)	Width (in.)	Depth (in.)
960	0.56	16	10	6
1111	0.64	17⅝	11	6⅝
1659	0.96	19¼	12	7¼

Table B-8. 8-inch Woofer — Two Way

Volume (cu. in.)	Volume (cu. ft.)	Height (in.)	Width (in.)	Depth (in.)
963	0.56	16⅛	10	6
1115	0.65	16⅞	10½	6⁵⁄₁₆
1281	0.74	17¹¹⁄₁₆	11	6⅝
1668	0.97	19⁵⁄₁₆	12	7¼

Table B-9. 8-inch Woofer — Three Way

Volume (cu. in.)	Volume (cu. ft.)	Height (in.)	Width (in.)	Depth (in.)
974	0.56	16⅛	10	6⅛
1295	0.75	17¹¹⁄₁₆	11	6¹¹⁄₁₆
1695	0.98	19⅜	12	7⁵⁄₁₆

Table B-10. 10-inch Woofer — One Way

Volume (cu. in.)	Volume (cu. ft.)	Height (in.)	Width (in.)	Depth (in.)
1674	0.96	19¼	12	7¼
1875	1.09	20	12½	7½
2109	1.22	20¾	13	7¾
2634	1.52	22⅜	14	8⅜

Table B-11. 10-inch Woofer — Two Way

Volume (cu. in.)	Volume (cu. ft.)	Height (in.)	Width (in.)	Depth (in.)
1663	0.96	19⁵⁄₁₆	12	7½
1880	1.09	20⅛	12½	7½
2114	1.22	20⅞	13	7¾
2646	1.53	22½	14	8⅜

Table B-12. 10-inch Woofer — Three Way

Volume (cu. in.)	Volume (cu. ft.)	Height (in.)	Width (in.)	Depth (in.)
1897	1.10	20⅛	12½	7⅝
2133	1.23	20⅞	13	7⅞
2979	1.72	23⅜	14½	8¾

Table B-13. 12-inch Woofer — One Way

Volume (cu. in.)	Volume (cu. ft.)	Height (in.)	Width (in.)	Depth (in.)
3240	1.88	24	15	9
3932	2.28	25 5/8	16	9 1/2
4716	2.73	27 1/4	17	10 1/4
5599	3.24	28 3/4	18	10 3/4

Q: bet. 1.0 & .8 PEAK! 1. dB

Table B-14. 12-inch Woofer — Two Way

Volume (cu. in.)	Volume (cu. ft.)	Height (in.)	Width (in.)	Depth (in.)
3247	1.88	24 1/8	15	9
3940	2.28	25 11/16	16	9 5/8
4725	2.73	27 5/16	17	10 1/4
5618	3.25	28 7/8	18	10 3/4

Table B-15. 12-inch Woofer — Three Way

Volume (cu. in.)	Volume (cu. ft.)	Height (in.)	Width (in.)	Depth (in.)
4757	2.75	25 5/16	17	10 5/16
5644	3.27	28 7/8	18	10 7/8
6675	3.86	30 5/8	19	11 1/2

Table B-16. 15-inch Woofer — One Way

Volume (cu. in.)	Volume (cu. ft.)	Height (in.)	Width (in.)	Depth (in.)
5599	3.24	28 3/4	18	10 3/4
7680	4.44	32	20	12
10222	5.92	35 1/4	22	13 1/4
13271	7.68	38 3/8	24	14 3/8

Table B-17. 15-inch Woofer — Two Way

Volume (cu. in.)	Volume (cu. ft.)	Height (in.)	Width (in.)	Depth (in.)
5608	3.25	28 7/8	18	10 3/4
7692	4.45	32 1/8	20	12
10237	5.92	35 5/16	22	13 1/4
13306	7.70	38 1/2	24	14 3/8

Table B-18. 15-inch Woofer — Three Way

Volume (cu. in.)	Volume (cu. ft.)	Height (in.)	Width (in.)	Depth (in.)
7736	4.48	32⅛	20	12⅛
10290	5.95	35⁵⁄₁₆	22	13⁵⁄₁₆
13415	7.76	38⅝	24	14½

METRIC CONVERSIONS

International System of Units (SI) — Metric Units

Prefix	Symbol	Multiplication Factor	
exa	E	10^{18}	= 1,000,000,000,000,000,000
peta	P	10^{15}	= 1,000,000,000,000,000
tera	T	10^{12}	= 1,000,000,000,000
giga	G	10^{9}	= 1,000,000,000
mega	M	10^{6}	= 1,000,000
kilo	k	10^{3}	= 1,000
hecto	h	10^{2}	= 100
deca	da	10^{1}	= 10
(unit)		10^{0}	= 1
deci	d	10^{-1}	= 0.1
centi	c	10^{-2}	= 0.01
milli	m	10^{-3}	= 0.001
micro	u	10^{-6}	= 0.000001
nano	n	10^{-9}	= 0.000000001
pico	p	10^{-12}	= 0.000000000001
femto	f	10^{-15}	= 0.000000000000001
atto	a	10^{-18}	= 0.000000000000000001

1 meter (m) =
100 centimeters (cm) =
1000 millimeters (mm)

25.4	mm	=	1 inch
2.54	cm	=	1 inch
30.48	cm	=	1 foot
0.3048	m	=	1 foot
0.9144	m	=	1 yard
1.609	km	=	1 mile
1.852	km	=	1 nautical mile

Fractional Dimensions

Inches	Millimeters	Inches	Millimeters	Inches	Millimeters
1/64	0.016	0.397	23/64	0.359	9.128	11/16	0.688	17.463
1/32	0.031	0.794	3/8	0.375	9.525	45/64	0.703	17.859
3/64	0.047	1.191	25/64	0.391	9.922	23/32	0.719	18.256
1/16	0.063	1.588	13/32	0.406	10.319	47/64	0.734	18.653
5/64	0.078	1.984	27/64	0.422	10.716	3/4	0.750	19.050
3/32	0.094	2.381	7/16	0.438	11.113	49/64	0.766	19.447
7/64	0.109	2.778	29/64	0.453	11.509	25/32	0.781	19.844
1/8	0.125	3.175	15/32	0.469	11.906	51/64	0.797	20.241
9/64	0.141	3.572	31/64	0.484	12.303	13/16	0.813	20.638
5/32	0.156	3.969	1/2	0.500	12.700	53/64	0.828	21.034
3/16	0.188	4.762	33/64	0.516	13.097	27/32	0.844	21.431
13/64	0.203	5.159	17/32	0.531	13.494	55/64	0.859	21.828
7/32	0.219	5.556	35/64	0.547	13.891	7/8	0.875	22.225
15/64	0.234	5.953	9/16	0.563	14.288	57/64	0.891	22.622
1/4	0.250	6.350	37/64	0.578	14.684	29/32	0.906	23.019
17/64	0.266	6.747	19/32	0.594	15.081	59/64	0.922	23.416
9/32	0.281	7.144	39/64	0.609	15.478	15/16	0.938	23.813
19/64	0.297	7.541	5/8	0.625	15.875	61/64	0.953	24.209
5/16	0.313	7.938	41/64	0.641	16.272	31/32	0.969	24.606
21/64	0.328	8.334	21/32	0.656	16.669	1.0	1.000	25.400
11/32	0.344	8.731	43/64	0.672	17.066			

GLOSSARY

acoustic feedback: A squealing sound when the output of an audio circuit is fed back in phase into the circuit's input.

acoustic fiberglass: Thin fiberglass material used as damping material inside speaker enclosures.

acoustic suspension: A speaker designed for, or used in, a sealed enclosure.

ac coupling: Coupling between electronic circuits that passes only alternating current and time varying signals, not direct current.

acoustics: The science or study of sound.

air suspension: An acoustic suspension speaker.

alternating current (ac): An electrical current that periodically changes in magnitude and direction.

ampere (A): The unit of measurement for electrical current in coulombs (6.25×10^{18} electrons) per second. There is one ampere in a circuit that has one ohm resistance when one volt is applied to the circuit. See Ohm's law.

amplifier: An electrical circuit designed to increase the current, voltage, or power of an applied signal.

amplitude: The relative strength (usually voltage) of a signal. Amplitude can be expressed as either a negative or positive number, depending on the signals being compared.

attenuation: The reduction, typically by some controlled amount, of an electrical signal.

audio frequency: The acoustic spectrum of human hearing, generally regarded to be between 20 Hz and 20,000 Hz.

baffle: A piece of wood inside an enclosure used to direct or block the movement of sound.

balance: Equal signal strength provided to both left and right stereo channels.

bandpass filter: An electric circuit designed to pass only middle frequencies. See also high-pass filter and low-pass filter.

basket: The metal frame of a speaker.

bass: The low end of the audio frequency spectrum: approximately 20 Hz to about 1000 Hz.

bass reflex: A ported reflex speaker enclosure.

battens: Small strips of wood placed inside a speaker to reinforce its mating corners or to provide a mounting surface for front and back panels.

bobbin: A paper, plastic, or metal cylinder around which is wound the wire that forms a speaker's voice coil. The bobbin is mechanically connected to the speaker cone.

capacitor (C): A device made up of two metallic plates separated by a dielectric (insulating material). Used to store electrical energy in the electrostatic field between the plates. It produces an impedance to an ac current.

channel: The left or right signals of a stereo audio system.

circuit: A complete path that allows electrical current from one terminal of a voltage source to the other terminal.

clipping: A distortion caused by cutting off the peaks of audio signals. Clipping usually occurs in the amplifier when its input signal is too high or when the volume control is turned up too high.

coaxial driver: A speaker that is composed of two individual voice coils and cones; used for reproduction of sounds in two segments of the sound spectrum. See also triaxial driver.

coloration: "Smearing" sounds by adding frequencies due to intermodulation distortion. More prevalent at high audio frequencies.

compliance: The relative stiffness of a speaker suspension, typically indicated simply as "high" or "low," but technically specified as Vas.

cone: The cone-shaped diaphragm of a speaker attached to the voice coil. It produces pulsations of air that the ear detects as sound.

crossover network: An electric circuit or network that splits the audio frequencies into different bands for application to individual speakers.

current (I): The flow of charge measured in amperes.

damping: 1. Acoustic fiberglass material used inside speaker enclosures. 2. The reduction of movement of a speaker cone, due either to the electromechanical characteristics of the speaker driver and suspension, or the effects of pressure inside a speaker enclosure.

decibel (dB): A logarithmic scale used to denote a change in the relative strength of an electric signal or acoustic wave. It is a standard unit for expressing the ratio between power level P_1 and power level P_2. dB = 10 $\log_{10} P_1/P_2$. An increase of 3 dB is a doubling of electrical (or signal) power; an increase of 10 dB is a doubling of perceived loudness. The decibel is not an absolute measurement, actually, but indicates the relationship or ratio between two signal levels.

direct current (dc): Current in only one direction.

dispersion: The spreading of sound waves as they leave a speaker.

distortion: Any undesirable change in the characteristics of an audio signal.

dome tweeter: A high frequency speaker with a dome-shaped diaphragm that provides much better dispersion of high frequencies than standard cone speakers.

driver: The electromagnetic components of a speaker, typically consisting of a magnet and voice coil.

ducted port: A ported reflex speaker enclosure.

dynamic-range: The range of sound levels which a system can reproduce without distortion.

dynamic range: The range of sounds, expressed in decibels, between the softest and loudest portions.

equalizer: An adjustable audio filter inserted in a circuit to divide and adjust its frequency response.

equalization: As used in audio, the adjustment of frequency response to tailor the sound to match personal preferences, room acoustics, and speaker enclosure design.

fader: A variable control used to change the distribution of power between front and rear speakers.

farad: The basic unit of capacitance. A capacitor has a value of one farad when it can store one coulomb of charge with one volt across it.

filter: An electrical circuit designed to prevent or reduce the passage of certain frequencies.

flat response: The faithful reproduction of an audio signal; specifically, variations in output level of less than one decibel above or below a median level over the audio spectrum.

free air resonance: The natural resonant frequency of a woofer speaker when operating outside an enclosure.

frequency: The number of waves (or cycles) arriving at or passing a point in one second; expressed in hertz (or Hz).

frequency response: The range of frequencies that are faithfully reproduced by a given speaker or audio system.

fundamental or fundamental tone: The tone produced by the lowest frequency component of an audio signal.

full-range: A speaker designed to reproduce all or most of the sound spectrum.

golden ratio: The ratio of the depth, width, and height of a speaker enclosure, based on the Greek Golden Rectangle, and which most often provides the best sound. W = 1.0, Depth = 0.6W, Height = 1.6W

grille cloth: Fabric used to cover the speaker mounted in an enclosure.

ground: Refers to a point of (usually) zero voltage, and can pertain to a power circuit or a signal circuit.

harmonic: The multiple frequencies of a given sound, created by the interaction of signal waveforms. A "middle C" on the piano has a fundamental audio frequency of 256 Hz, but also a number of secondary higher frequencies (harmonics) that are odd and even multiples of this fundamental.

harmonic distortion: Harmonics artificially added by an electrical circuit or speaker, and are generally undesirable. It is expressed as a percentage of the original signal.

hertz: A unit of frequency equal to one cycle per second, named after German physicist H.R. Hertz.

high-fidelity: Commonly called hi-fi, it refers to the reproduction of sound with little or no distortion.

high-pass filter: An electric circuit designed to pass only high frequencies. See also bandpass filter and low-pass filter.

hiss: Audio noise that sounds like air escaping from a tire.

horn: A speaker design using its own funnel-shaped conduit to amplify, disperse, or modify the sounds generated by the internal diaphragm of the speaker.

hum: Audio noise that has a steady low frequency pitch, typically caused by the effects of induction by nearby ac lines or leakage of ac line frequency into an amplifier's signal circuits.

impedance: The opposition of a circuit or speaker to an alternating current.

inductance (L): The capability of a coil to store energy in a magnetic field surrounding it. It produces an impedance to an ac current.

L-pad: A type of potentiometer that maintains constant impedance at its input while varying the signal level at its output. L-pads are most often used as an external balance control or variable attenuator (volume control).

low-pass filter: An electric circuit designed to pass only low frequencies. See also bandpass filter and high-pass filter.

midrange: A speaker designed to reproduce the middle frequencies of the sound spectrum, generally most efficient between about 1000 Hz to 4000 Hz.

mounting flange: The outer edges of a speaker frame which has pre-drilled holes to accept screws or bolts for securing it to the enclosure.

noise: An unwanted sound.

ohm (Ω): A unit of electrical resistance or impedance.

Ohm's law: A basic law of electric circuits. It states that the current I in amperes in a circuit is equal to the voltage E in volts divided by the resistance R in Ohms; thus, I = E/R.

passive radiator (or drone): A speaker with a cone but no driver components. The cone vibrates with the change in pressure inside the speaker enclosure. Typically used to increase bass output with no increase in electrical power.

peak: The maximum amplitude of a voltage or current.

piezoelectric: A characteristic of some materials, especially crystal, that when subjected to electric voltage the material vibrates. Sometimes used in tweeters in place of a magnet, voice coil, and cone.

polarity: The orientation of magnetic or electric fields. The polarity of the incoming audio signal determines the direction of movement of the speaker cone.

ported reflex: A type of speaker enclosure that uses a duct or port to improve efficiency at low frequencies.

power: The time rate of doing work or the rate at which energy is used. A watt of electrical power is the use of one joule of energy per second. Watts of electrical energy equals volts times amperes.

resonance: The tendency of a speaker to vibrate most at a particular frequency; sometimes referred to as natural frequency.

resistance: In electrical or electronic circuits, a characteristic of a material that opposes the flow of electrons. It results in loss of energy in a circuit dissipated as heat. Speakers have resistance that opposes current.

RMS: An acronym for root mean square. The RMS value of an alternating current produces the same heating effect in a circuit as the same value of a direct current.

signal: The desired portion of electrical information.

signal-to-noise (S/N): The ratio, expressed in dB, between the signal (sound you want) and noise (sound you don't want).

sine wave: The waveform of a pure alternating current or voltage. It deviates about a zero point to a positive value and a negative value. Audio signals are sine waves or combinations of sine waves.

sound pressure level (SPL): The loudness of an acoustic wave stated in dB that is proportional to the logarithm of its intensity.

spider: The flexible fabric that supports the bobbin, voice coil, and inside portion of the cone within the speaker frame.

surround: The outer suspension of a speaker cone; the surround connects the outside portion of the cone to the speaker frame.

suspension: See surround.

three-way: A type of speaker system composed of three ranges of speakers, specifically a tweeter, midrange, and woofer. See also two-way.

total harmonic distortion (THD): The percentage, in relation to a pure input signal, of harmonically derived frequencies introduced in the sound reproducing circuitry and hi-fi equipment (including the speakers).

treble: The upper end of the audio spectrum, usually reproduced by a tweeter.

transient response: The instantaneous change in an electronic circuit's output response when input circuit conditions suddenly change from one steady-state condition to another.

triaxial driver: A speaker that is composed of three individual voice coils and cones; used for the reproduction of sounds in three segments of the sound spectrum. See also coaxial driver.

tweeter: A speaker designed to reproduce the high or treble range of the sound spectrum, generally most efficient from about 4000 Hz to 20,000 Hz.

two-way: A type of speaker system composed of two ranges of speakers, consisting of any two of the following: a tweeter, midrange, and woofer. See also three-way. Some midrange speakers are classified as midrange/ tweeter.

voice coil: The wire wound around the speaker bobbin. The bobbin is mechanically connected to the speaker cone and causes the cone to vibrate in response to the audio current in the voice coil.

watt: A unit of electrical power.

whizzer: A small supplementary cone attached to the center of the speaker's main cone for the purpose of increasing high frequency response.

woofer: A speaker designed to reproduce the low frequencies of the sound spectrum, generally most efficient from about 20 Hz to 1000 Hz.

INDEX